T0091597

9
F
Fluorine
77
Ir
Iridium
16
S
Sulfur
T
Temperature
1
D
Deuterium
8
O
Oxygen
102
No
Nobelium
1
H
Hydrogen
18
Ar
Argon
m
Mass

# First Do No Harm

## A Chemist's Guide to Molecular Design for Reduced Hazard

**Predrag V. Petrovic**
**Paul T. Anastas**

The term "First Do No Harm" with regard to molecular design was introduced by Nicholas D. Anastas, Ph.D., in his work "A Framework for Hazard Reduction through Green Chemistry," doctoral dissertation, University of Massachusetts Boston, 2008

JENNY STANFORD
PUBLISHING

*Published by*

Jenny Stanford Publishing Pte. Ltd.
101 Thomson Road
#06-01, United Square
Singapore 307591

Email: editorial@jennystanford.com
Web: www.jennystanford.com

**British Library Cataloguing-in-Publication Data**
A catalogue record for this book is available from the British Library.

**First Do No Harm: A Chemist's Guide to Molecular Design for Reduced Hazard**



ISBN 978-981-4968-59-1 (Hardcover)
ISBN 978-1-003-35964-7 (eBook)

*To my loving wife, Ana, who supported,*
*encouraged, and believed in me all the way,*
*even when I was doubting myself.*

Predrag V. Petrovic

*To the three people who bring love into my life,*
*Julie, Kennedy, and Aquinnah.*

Paul T. Anastas

*To every individual whose life was impacted or*
*ended due to a chemical hazard and to every*
*green chemistry scientist who is working to make*
*sure that this doesn't happen in the future.*

PVP and PTA

# Contents

# Preface

When green chemistry was first introduced as a new conceptual construct in 1991, it was carefully presented in a way that was not condemning of any chemistry that came previously. This was largely a strategic decision since it is always good practice to not insult the people you need to persuade to bring about needed change. To say, "The prior ways that we have pursued chemistry were flawed, foolish, and downright inelegant" would have simply ensured that no practicing chemist alive would want to be associated with green chemistry. It also would have been unfair. Because while it is trivially easy (as well as intellectually lazy) to condemn those who may have taken actions in relative ignorance in the past, it is far more important to look at our actions in a clear-eyed way *today* to see what is it that people in the future may correctly condemn about what we are doing *currently*.

While there was very little resistance to the concept of green chemistry since it was presented as simply a way to pursue the never-ending march of innovation in the chemical enterprise, there were those prideful practitioners of a bygone age that would protest by saying, "We don't need to design chemicals and chemical transformations to reduce the hazard. We are the best of the world at handling and managing hazardous substances and they are simply the tools of our trade." Most of these types of objections were easily disposed of by a simple thought experiment that asks, "If you have two chemicals, A and B, that are identical in every way; functional performance, cost, availability, etc., but one is completely innocuous and the other is acutely hazardous, which one would you choose?" Given all of the downsides of the hazardous alternative, the rational answer was that the innocuous option would be best, and therefore, should be at least *one consideration* in the design and choice of chemicals.

Now that it was understood that this is a reasonable goal to pursue in its own right, the question is, "How do we do it?" To a first approximation, chemistry and toxicology, have been disparate fields that have had little overlap through most of their existence until that latter part of the 20th century. Still today, a tiny sliver of chemistry departments in universities worldwide requires any training in molecular toxicology disregarding the fact that this would provide scientific insight into the *human and global consequences* of the tools of our trade. And yet, research communities have emerged over the decades and pursued molecular insights of biological activity and borrowed from the magnificent lessons of the pharmaceutical industry and gleaned from the lessons of the pesticide industry and learned from the studies conducted to protect the environment. Although zero adverse consequence chemicals will never be achieved, we must keep zero adverse consequences as our goal of perfection because it is the pursuit of that unattainable goal that will guarantee continuous improvement.

Today, there are conceptual frameworks and a knowledge base that is important for every scientist who chooses to design a molecule or material. This is a dynamic field and the knowledge and concepts will continue to grow and be refined. But at this stage in history, to proceed with introducing new substances into the world that it has never seen without the rigorous design consideration of how to minimize adverse consequences, would border on the unethical. So, chemist, this volume is offered for your use as you pursue new inventions and innovations to enable you to **First Do No Harm**.*

**Predrag V. Petrovic**
**Paul T. Anastas**
November 2022

*Anastas, N. D. A Framework for Hazard Reduction through Green Chemistry, Ph.D. dissertation, University of Massachusetts Boston, 2008.

# Acknowledgments

First and foremost, we would like to thank Tobias Muellers, a master of environmental science at the Yale School of the Environment, for his significant contribution in the making of this book. He participated in all stages of the production of this manuscript by being involved in the discussions on the contents of the chapters, working on the preparation of figures and tables, organizing the references, proofreading, and additionally contributing to several case studies in Chapter 8.

This book would not have been possible without the foundational work on designing safer chemicals in the Center for Green Chemistry and Green Engineering at Yale laid by Julie B. Zimmerman, Adelina Voutchkova-Kostal, Jakub Kostal, and Philip Coish. We would also like to thank Julie B. Zimmerman, Evan Beach, and Momoko Ishii for reading the manuscript and providing valuable feedback that helped improve the content. A special thank you is dedicated to all of the past and current members of the center who have, through the years, contributed to the advancement of the center's missions.

Finally, we would like to thank all the staff at Jenny Stanford Publishing for all their contributions in editing, arranging, and publishing this book.

**Predrag V. Petrovic**

**Paul T. Anastas**

# Introduction

Two hundred years ago, the idea that a person could combine ingredients that could not be seen and create structures and substances that had never existed before would be considered the scientific equivalent of a miracle. However, that is exactly what synthetic chemists have been doing for about two centuries. Chemists started making molecules long before they knew what a molecule was. Friedrich Wöhler synthesized urea in 1828,[1] almost 100 years before Bohr would even propose a simple model for what the atom was. William Perkin was already commercially producing the first synthetic chemical—aniline derived dye[2]—for three years before Mendeleev even proposed the periodic table of the elements. It is no wonder that synthetic chemistry has had something of a hint of being a super-power, and that some historical giants of organic chemistry have considered themselves near-gods. The feats of chemistry, especially in the 19th and the first half of the 20th century, have indeed bordered on the divine; however, this is definitely not a compliment.

This is because chemists are scientists, and divinity has nothing to do with the scientific approach that seeks to understand the underlying truths of nature by applying the scientific method. This procedure relies on making observations, forming hypotheses, and designing experiments to devise laws and theories. Being able to accomplish things in the absence of knowledge, insight, and intention is at most fortuitous. It may even be profitable and world-changing, but for scientists, it is not satisfactory. And thus, synthetic chemists were never satisfied, and as each new achievement of synthesizing a new molecule was attained, it was inextricably coupled to a fervent desire to understand how and why the synthesis was possible. More specifically, the progress in the field was driven by numerous intuitive questions such as:

- What was the fundamental structure of the molecule?
- What was the nature of the chemical bond?
- What conditions allowed the synthesis to happen?

- What was the mechanism and why was one pathway favored over another?
- What were the properties of the starting material that allowed for the transformation?
- What were the properties of the product that resulted in its stability?
- What were the properties of the transition state that were essential to the synthetic pathway?
- And most importantly, how could we **control** the whole process?

That was the key—**control**. In order to truly master a system, you had to control it. And so synthetic chemists relentlessly pursued control until many would be able to confidently state that virtually every molecular structure that can be drawn can be synthesized. Control was explored with different forms—regioselective control, stereoselective control, and enantiomeric control. And why was control so important? Because only with control can you **design.**

Design is a statement of human intention; one cannot design by accident.[3] If it is an accident, it is certainly not designed. As much as Wöhler and Perkin were pioneers, their most famous achievements, while appropriately hailed as pioneering breakthroughs, were not by design. They were wonderfully serendipitous yet unintentional.

What transpired over the years throughout the 20th century was the design of chemicals with a particular function or purpose in mind. Whether it was medicine, paint, or plastic, all too often we found out that these wonderfully functional chemicals had problems of toxicity and caused hazard.

This was a result of a tremendous amount of ignorance about what caused adverse effects based on a chemical's nature; in fact, we had almost no insight. The reason that our lack of insight was so important is that a lot of regulations, laws, and policies were written without proper understanding of the underlying causes of hazard. Since then, science has evolved very quickly and deeply in so many different ways that have empowered us and allowed us to think differently about molecular design.

Pharmacokinetics serve as one illustrative example. How we get exposed to chemicals and how chemicals get into us depends on four processes: absorption, distribution, metabolism, and excretion (ADME). The different ways that AMDE manifests is directly relevant to pharmaceutical drugs, and how they are made. For over 50 years, the pharmaceutical industry has been trying to figure out how to make a molecule that is stable during manufacturing and packaging, stable when it is on the shelf, and stable in a medicine cabinet until the pill is swallowed. And then, when the pharmaceutical drug gets into the body it must cross the necessary biological membranes and barriers, bind to a particular receptor in order to cause its therapeutic effect, and finally be excreted into the environment. However, this last step in the process, i.e., that the drug gets excreted *harmlessly* into the environment, has not really been part of the goals for drug discovery and development. Basically, green chemistry and molecular design for reduced hazard are taking all of that knowledge and turning it on its head, i.e., thinking of the problem the other way around, and considering full implications.[3] In other words, the questions arise for how to design a molecule such that:

- it never gets into the body (i.e., is never able to be ingested or inhaled or transported through the skin)
- is never able to cross the biological membranes
- is never able to bind to receptors
- is never able to cause a toxic effect

Thus, all of the insight and wisdom that has been developed by the pharmaceutical industry is converted and adapted for safety purposes. In other words, today we can design molecules that are not biologically active, rather than biologically active molecules as desired for pharmaceuticals.

The essence of this strategy is to reduce bioavailability. To ensure this, the designer must ensure that the molecule is not able to be distributed around the body and that it is not able to bind to receptors. Additionally, if the molecule does get into the body, excretion potential should be increased. One of the main factors in these considerations is the lipid bilayer, the barrier that is a primary component of most of the membranes in our bodies. Understanding the lipid bilayer is so important because it is the gateway for molecules to get into the body through a variety of barriers, such as the dermis, the epithelium, and different organs. Pharmaceutical knowledge in the areas of crossing the lipid bilayer and absorption already provides information for the chemical design.

Absorption is the primary way that a chemical gets into the body through inhalation, ingestion, or even entering through dermal exposures, and is closely related to distribution. The extent of absorption determines the dose and the bioavailability. In pharmacology, bioavailability is defined as a category of absorption, i.e., it is the fraction of the drug that reaches the systemic circulation, while absorption represents the movement of the drug across the outer mucosal membranes of the GI tract.[4] The concept of absorption was carefully investigated by Chris Lipinski, a pioneer who had been working at Pfizer for 30 years. Lipinski wanted to increase the bioavailability of pharmaceuticals because at that time most developed drugs failed phase three clinical trials due to a lack of bioavailability. As such, this was one of the great challenges in the pharmaceutical industry. Thus, he came up with a set of design guides, now known as Lipinski's rule of five.[5] This set of rules contains molecular properties that a drug molecule should possess to be bioavailable. These guides will be covered in more detail in later chapters.

Building on the rules of Chris Lipinski, we are now looking at the same process of molecular design to limit absorption across the membranes by focusing on those specific physical-chemical properties that are going to either enhance or disfavor absorption into the body, bioavailability, and distribution of the chemical (Fig. 1).

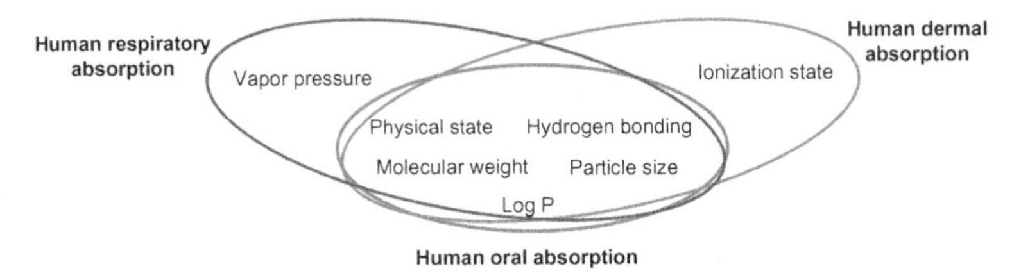


**Figure 1** Molecular properties involved in human respiratory, oral, and dermal absorption.

One of the factors that are tremendously important in absorption in the digestive tract is particle size. Particle size of more than 100 nanometers is desirable from a molecular design perspective, and if a particle is smaller, that should raise a red flag.[6] Another factor, the oil/ water partition coefficient (log P), should be either less than zero or more than five.[6] If it falls in those windows, then absorption is going to be disfavored. Where it falls between zero and five, that is going to be of tremendous concern, especially in the digestive tract. If there is a choice between something being in a solid phase versus a liquid phase, a solid should be preferred. Additionally, increasing the number of hydrogen bonds that can be formed by the molecule is something that is greatly preferred. Something as simple as molecular weight ($M_w$), can also influence the absorption. A $M_w$ greater than 500 Dalton's atomic mass units is going to be significantly better since these molecules will be absorbed less across the GI tract.[6] At the same time, the window of the permeability for log P will prevent molecules that are either extremely water-soluble or extremely fat-soluble from making their way across the lipid bilayer which can be utilized for safer molecular design.[5,6]

Many of the same basic physicochemical properties influence absorption in the respiratory tract. If the chemical is larger than five nanometers, it is not going to be respired deep into the lungs, whereas if it is smaller than five nanometers, it is going to be very bioavailable.[6] In this case, log P values are again less than zero and greater than five, while the molecular weight cutoff is 400 Daltons.[6] Finally, a new property—vapor pressure—becomes very important. Ideally, the molecule should not be respired at all, but more realistically vapor pressure needs to be less than 0.001 mmHg.[6]

With dermal adsorption, the same types of properties are again relevant. Because of the structure of the epidermis and dermis molecules with log P values between zero and six and $M_w$ above 500 Dalton's are less likely to be absorbed.[6] But, in this case, physical state is extremely important. As we know intuitively, solids are far less likely to be transported than liquids. Additionally, ionization is also relevant since the ability of a substance to be ionized and to have a charge is going to cause molecules to be extremely unlikely to be able to be transported across the dermal layers.

If, or more likely when, a chemical gets into the body, another process from the ADME list becomes relevant (Fig. 2). Biotransformation, or more specifically metabolism, is a process able to transform a molecule from its parent state to a derivative or derivatives, very often through oxidation. This process aims to detoxify hazardous chemicals. However, this can also result in a molecule that will be more toxic than the parent molecule. While these processes can be very complicated and will be explored in more detail later, it is important to emphasize that detoxification can be promoted by careful molecular design. Certain functional groups, such as esters or amines, are key structural features that influence whether the molecule will be activated and become more toxic or deactivated and less toxic through metabolism.[6] For example, ester groups are prone to hydrolysis which is believed to be a helpful mechanism to eliminate exogenously ingested esters that could be harmful.[7] The oxidation of primary and secondary amines into hydroxylamines was found to be the cause of hepatotoxic and mutagenic effects.[8]

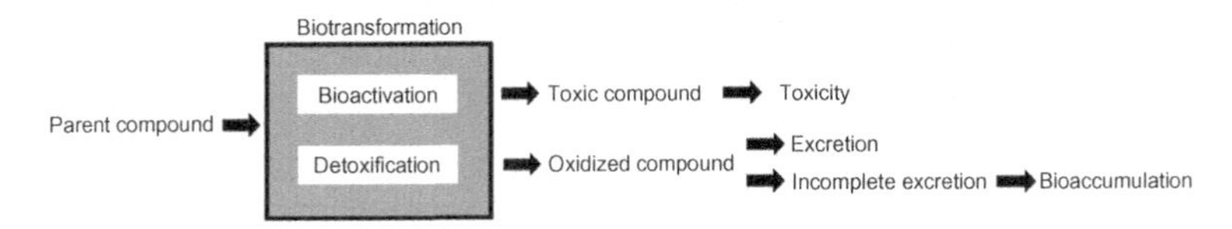

**Figure 2** General overview of metabolism and possible consequences.

In its parent state, a molecule may not nearly be as much of a concern, but after it is metabolized and goes through biotransformation and/or bioactivation, resulting metabolite(s) can become toxic (e.g., nonylphenol ethoxylates).[9] This is where the consequences related to liver, lung, and kidney cancer come about since these organs are responsible for detoxifying processes in the body. Metabolism is the natural way to detoxify harmful chemicals in the processes where the molecule, after being oxidized, becomes far more susceptible to excretion. These metabolic processes make molecules more water-soluble and more likely to pass through the body without being able to be distributed and interact with other organs.

Nonylphenol ethoxylates (NPEs) are nonionic surfactants (detergent-like substances) used for industrial processes and in consumer laundry detergents, personal hygiene, automotive, latex paints, and lawn-care products. Once released in the environment, they break down into highly persistent, bioaccumulative, and toxic nonylphenol.

One infamous example of a molecule that caused significant environmental harm and sparked environmental activism in the 1960s and more careful thinking on the fate of chemicals is the case of insecticide dichlorodiphenyltrichloroethane (DDT) (Fig. 3). While it was used for the noble purpose of trying to defeat malaria, typhus, and bubonic plague by trying to kill pests that were spreading disease, its use caused significant problems in wildlife.[10] This molecule cannot be broken down by metabolic transformations, making it very persistent. Because of its properties, it is instead gathered in the cells which resulted in the bioaccumulation and biomagnification of DDT across the food chain, and ultimately the death of many avian species due to interference with nerve impulses.[11]

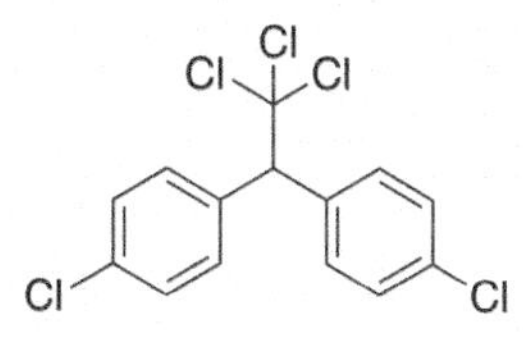


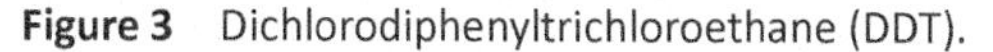

**Figure 3** Dichlorodiphenyltrichloroethane (DDT).

While the previous example left long-lasting consequences for the environment, it is a hardly isolated case of pursuing chemicals with specific functions without forethought about its environmental fate. However, vast experience gained through pharmaceutical drug development provides an understanding of the underlying processes and properties that guide the behavior of chemicals once they get into the body and the environment. Thus, it should be possible to identify a small subset of physicochemical properties to correlate both fundamental toxicokinetic (description of how the body interacts with a chemical as a function of dose and time within ADME framework) and toxicodynamic (the molecular, physiological, and biological effects of chemicals or their metabolites as a result of interaction with biological systems) susceptibility, and to specify those toxic endpoints (Fig. 4).

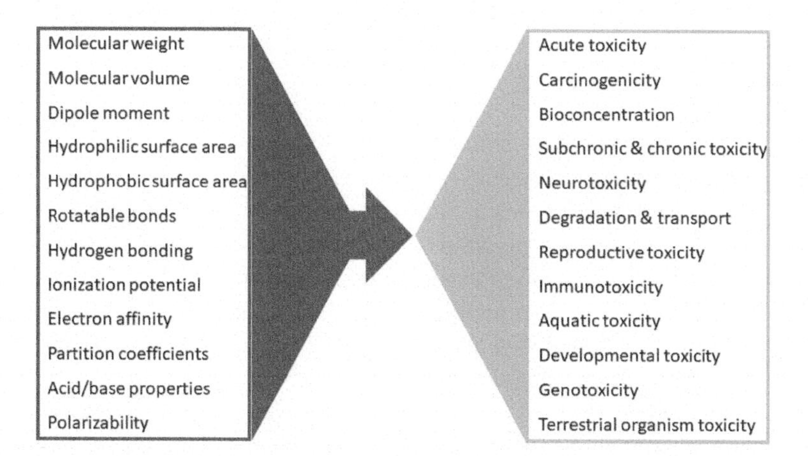


**Figure 4** Correlation of physicochemical properties and toxic endpoints.

This knowledge can be used to devise basic guidelines on how to design safer chemicals and reveal the pathways towards a framework for safer chemical design. This finally leads us to the concept of the molecular design pyramid shown in Fig. 5, which provides an overview of processes and properties playing a crucial role in safer chemical design.

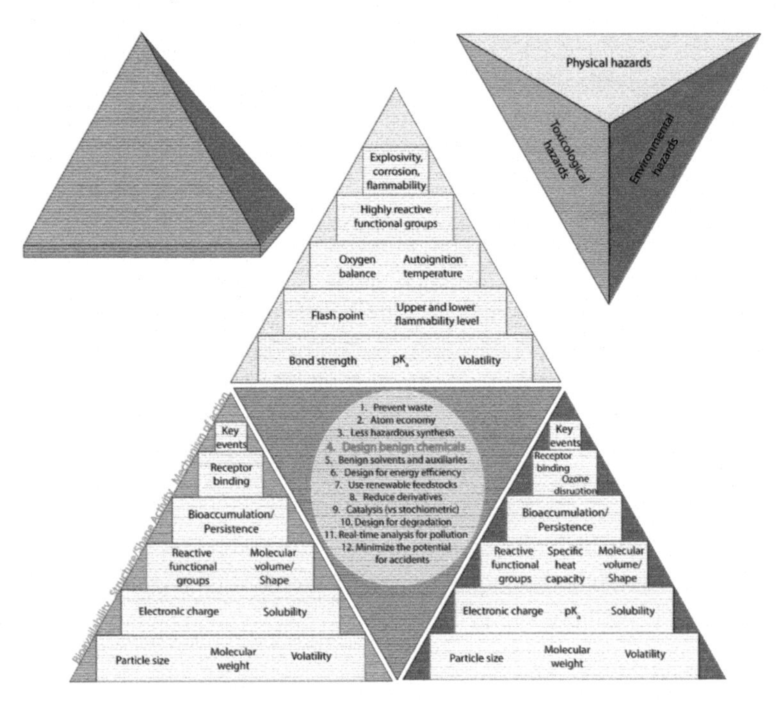


**Figure 5** Molecular design pyramid.

This pyramid is a symbolic representation of the factors, i.e., physicochemical properties and related processes, that have an impact on different types of hazards that molecules can cause. Each side of the unfolded pyramid is defined by one type of hazard—toxicological, environmental, and physical hazard—while at the base of the pyramid lie 12 principles of green chemistry[12] providing the resources and knowledge to connect each of the sides of the pyramid with emphasis on safer chemical design.

The most studied and understood group of properties and processes are shown on the toxicological hazard face of the pyramid since they are most relevant to human health. At the base of the pyramid lie molecular properties that influence bioavailability and govern the toxicokinetics of a molecule's interaction with biological systems. Often, these properties can be manipulated relatively easily, making them the simplest tool in the molecular design portfolio to eliminate or reduce bioavailability. Properties such as molecular weight and size, volatility, electronic charge, and solubility (amongst others), influence the absorption across the different biological membranes along with the distribution and transport of the molecules through the body. Through modification of these properties, the ability for chemicals to access the insides of cells and tissues, where they can cause various adverse effects, can be carefully governed.

Higher up the pyramid, more complex factors such as shape and structure affect the activity of chemicals. Most chemical reactions that are performed daily in the countless laboratories across the world involve transformations of molecules that range from very simple (e.g., simple double displacement reaction, $H_2SO_4 + 2\ NaOH \rightarrow Na_2SO_4 + 2\ H_2O$) to extremely complex (e.g., total synthesis,[13] processes of complete synthesis of complex molecules, such as natural products, from simple precursors). These transformations comprise changes to the parts of the molecule that often define its function and govern its reactivity, i.e., to the molecule's functional groups. This fact reveals a powerful tool—modifications of a (reactive) functional group, which was traditionally used to design chemicals for desired functionality, can instead be used in thoughtful chemical design of chemicals with their whole life cycle in mind. Too often we have created chemicals that perform their function with high efficiency but cause significant adverse effects, to living organisms in particular, and the environment as a whole (think of DDT, asbestos, plastics, etc.). A number of these man-made chemicals are very persistent, bioaccumulative, and toxic.

Finally, at the top of the pyramid, the most intricate factors influence the mechanisms of action of chemicals that spans key events in the cell and interactions with biomacromolecules. And the upper blocks of the pyramid are laid on top of the lower blocks, such that the more complex factors are built upon and are influenced by the simpler attributes at the molecular level. Thus, we have to be thoughtful of the potential unwanted interactions of our chemicals with the various targets in the cells, including receptors, enzymes, other proteins, DNA/RNA, etc. While some chemicals (pharmaceuticals) are intended to interact with such targets to cause positive effects, the design of all other chemicals that are not intended to be bioactive should carefully consider the chemical mechanism of action, and avoid known (and try to anticipate unknown) unwanted processes.

The second face of the pyramid defines the properties responsible for causing environmental hazards. As we will see later, environmental hazards can have ecotoxicological impacts affecting whole ecosystems, but also significantly impact the whole planet, making them global hazards. While the ecotoxicological hazards share the same group of properties, as shown on the toxicological face of the pyramid, chemicals causing global effects have some additional properties that define their hazard potential. These properties, such as ozone disruption potential, specific heat capacity, and $pK_a$ can directly influence climate change or the formation of ozone holes and consequently cause a hazard on a global scale.

Finally, the third face of the pyramid contains the properties related to the physical hazard. This group of properties is more distinct from the other two faces of the pyramid. They describe chemical properties and processes that can result in physical harm or damage. Many of these properties relate directly to the strength of the bonds within the molecule represented at the base of the pyramid. In other words, we can say that the energy stored in chemical bonds can get (often violently) released in a form of intense heat (flame) or as a shockwave (explosion). These processes stand at the top of the pyramid as a hazardous mechanism of action. In between are properties that were derived for clearer categorization of physical hazards. Properties such as lower and upper flammability level, flash point, oxygen concentration, and autoignition temperature directly influence the flammability of a substance; explosivity is affected by oxygen balance (OB) and the presence of very reactive functional groups; corrosive chemicals are defined by their $pK_a$ values. The broad range of chemicals that are covered by this category are often very reactive, be it that they are flammable, explosive, corrosive, or act as oxidizers.

Coming back to those questions posed by synthetic chemists who pursue safer chemical design at the beginning of their synthesis process, we can now give unambiguous answers that can in turn be used for rational and intentional design. Our understanding of the properties that influence the behavior of chemicals reveals the incredible potential that green chemistry has in using existing research about fundamental questions, and, as we will see in the following chapters, applying the gained knowledge in modification of these properties directly leads to designing safer chemicals.

## References


1. Kinne-Saffran, E.; Kinne, R. K. H. Vitalism and synthesis of urea. *Am J Nephrol*, 1999, **19**(2), 290–294.
2. Cova, T. F. G. G.; Pais, A. A. C. C.; Seixas de Melo, J. S. Reconstructing the historical synthesis of mauveine from Perkin and Caro: procedure and details. *Sci Rep*, 2017, **7**(1), 6806.
3. Anastas, P. T.; Zimmerman, J. B. The periodic table of the elements of green and sustainable chemistry. *Green Chem*, 2019, **21**(24), 6545–6566.
4. Chiou, W. L. The rate and extent of oral bioavailability versus the rate and extent of oral absorption: clarification and recommendation of terminology. *J Pharmacokinet Pharmacodyn*, 2001, **28**(1), 3–6.
5. Lipinski, C. A.; Lombardo, F.; Dominy, B. W.; Feeney, P. J. Experimental and computational approaches to estimate solubility and permeability in drug discovery and development settings. *Adv Drug Deliver Rev*, 1997, **23**(1–3), 3–25.
6. Voutchkova, A. M.; Osimitz, T. G.; Anastas, P. T. Toward a comprehensive molecular design framework for reduced hazard. *Chem Rev*, 2010, **110**(10), 5845–5882.
7. Laizure, S. C.; Herring, V.; Hu, Z.; Witbrodt, K.; et al. The role of human carboxylesterases in drug metabolism: have we overlooked their importance? *Pharmacotherapy*, 2013, **33**(2), 210–222.
8. Macherey, A.-C.; Dansette, P. M. Chapter 33: Biotransformations leading to toxic metabolites. In: Wermuth, C. G. (Ed.), *The Practice of Medicinal Chemistry*, 3rd ed., Academic Press, 2008, 674–696.
9. EPA. Certain nonylphenols and nonylphenol ethoxylates; significant new use rules, *Fed Regist*, 2014–23253, 59186–59195.
10. IPCS. Ultraviolet radiation. *Environ Health Criteria 160*, WHO, 1994, ISBN 92-4-157160-8.
11. Klaassen, C. D.; Watkins, III, J. B. *Casarett & Doull's Essentials of Toxicology*, 3rd ed., McGraw-Hill Education: New York, 2015.
12. Anastas, P. T.; Warner, J. C. *Green Chemistry: Theory and Practice*. Oxford University Press, 1998.
13. Nicolaou, K. C.; Vourloumis, D.; Winssinger, N.; Baran, P. S. The art and science of total synthesis at the dawn of the twenty-first century. *Angew Chem Int Ed*, 2000, **39**(1), 44–122.

# Chapter 1

# Hazard

## 1.1 What Is a Chemical Hazard?

A chemical hazard represents the potential of a chemical substance to cause harm to a person or the environment upon exposure due to its inherent properties. Chemical hazards are categorized by the type of hazard they pose as health, environmental, and physical hazards; however, some chemicals can also be hazardous in more than one way, which really depends on that particular chemical. These categories are specifically defined in the Globally Harmonized System of Classification (GHS) and Labelling of Chemicals document published by the United Nations Economic Commission for Europe (UNECE),[1] and U.S. Occupational Safety and Health Administration (OSHA) standard 1910.1200.[2] In the United States, OSHA requires that the information about chemical hazards be provided in Material Safety Data Sheets (MSDSs) and labels (Fig. 1.1) following the guidelines defined by UNECE.

Here, we will briefly describe each category and then expand on each in the following sections.

*First Do No Harm: A Chemist's Guide to Molecular Design for Reduced Hazard*
Predrag V. Petrovic and Paul T. Anastas

ISBN 978-981-4968-59-1 (Hardcover), 978-1-003-35964-7 (eBook)
www.jennystanford.com

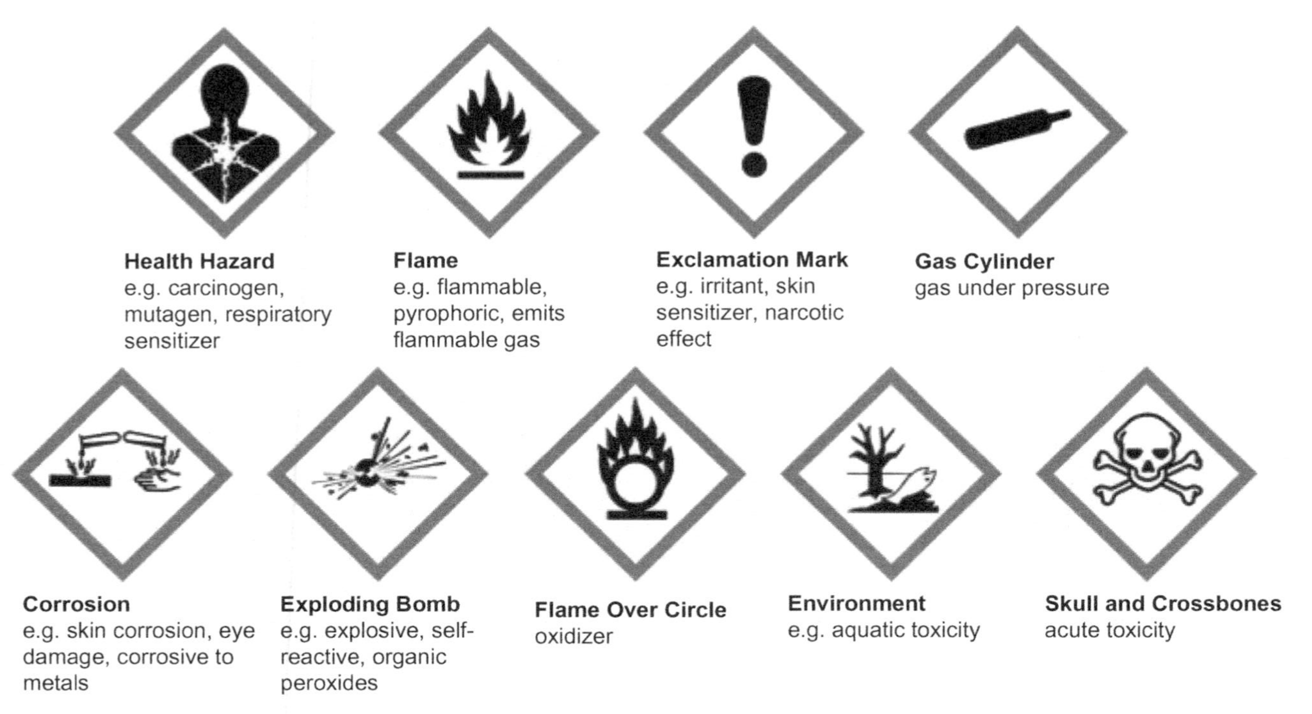


**Figure 1.1** Pictograms used in the GHS labels for chemical hazard categories.[3]

Health hazards are caused by chemicals that exhibit acute toxicity via any route of exposure and can be manifested as specific organ toxicity, carcinogenicity, reproductive toxicity, germ cell mutagenicity, respiratory or skin sensitization, skin corrosion or irritation, eye damage or irritation, and aspiration hazards. Since the body is a very complicated system finely tuned to function in an optimal fashion, large numbers of chemicals can disrupt this balance causing harm to the health of an individual. This can be done in various ways, differing in the amount of time it takes for the disruption to happen, the exact mechanism of action, the route chemical takes, the harm it causes, and so on. For example, with acutely toxic substances the effect is exhibited in a short amount of time (up to 24 hours) which includes gastrointestinal damage caused by arsenic, respiratory failure caused by mercury or chlorine, and serious eye damage caused by 2-mercaptoethanol or phenol, amongst many others. Carcinogenic substances (e.g., cadmium, benzene, vinyl chloride) cause adverse effects chronically, usually over a long period of time taking months or even years to exhibit.[4] Sensitizers such as formaldehyde lead to an inflammatory skin reaction or respiratory disorders such as asthma, upon repeated exposure to the chemical. Mutagenic chemicals (e.g., ethyl methanesulfonate, bromine) can cause or increase the frequency of mutations in an organism.[4]

Environmental hazards, as defined by the GHS, are generated by chemical substances causing significant damage to the environment and organisms living in it.[1] The origin of most substances considered to be in this group is anthropogenic, but some naturally occurring substances (e.g., lead, radon) can be environmental hazards as well. The main concerns that this category covers are two-fold; there are chemicals that cause global effects (e.g., climate change, ozone layer holes) and those that cause ecotoxicological acute and chronic effects on aquatic, sediment, or terrestrial life forms. Short-term impacts exhibited by different chemicals can cause acute lethality to a number of species across the environment. On the other hand, long-term impacts include bioaccumulation that may, or may not have a toxic effect on some organisms, but pose a problem further up the food chain (e.g., DDT, mercury, cadmium); and degradation where some of the chemicals persist in the environment impacting the spawning rates, causing genetic problems in offspring and behavioral changes (e.g., PFOS, PCBs).

Finally, the physical hazards category consists of chemicals that can cause physical damage because they are flammable, corrosive, explosive, or act as oxidizers, including gases under pressure. Flammables are chemicals that are capable of burning or igniting and causing fire or combustion. These can be flammable gases (hydrogen, methane, ethylene, butane), liquids (acetone, toluene, diethyl ether), and solids (metallic sodium, metallic hydrides). Corrosives are chemicals that cause irreversible alterations or visible destruction to materials or living tissue. These chemicals begin to cause damage as soon as they get into contact with the affected material and can come in a form of gas, liquid, and solid. Most corrosives are either acids (hydrochloric, sulfuric, perchloric) or bases (ammonium hydroxide, potassium hydroxide, sodium hydroxide), but other chemicals can be corrosive as well (nitrogen dioxide, phenol). Explosives and oxidizers are chemicals that are highly reactive which usually results in violent reaction or explosion under ambient conditions, or when they come into contact with oxygen, water, or other chemicals. These chemicals can be oxidizers (nitric acid), air reactive/pyrophoric (*t*-butyl lithium, silane), water-reactive (metallic sodium, potassium, magnesium, barium), or explosives (TNT, benzoyl peroxide, picric acid).

## 1.2 What Is a Toxicological Hazard?

> *All substances are poisons, there is none that is not a poison. The right dose differentiates a poison and a remedy.*
>
> —Paracelsus, 16th century[5]

This quote by 16th-century alchemist Paracelsus reveals the fundamental concept in toxicology, that all chemicals, regardless of the source (natural or manufactured) are potentially toxic at some dose. This concept becomes important when the assessment of risk from the use of chemicals is considered. The amount of the substance causing it to become toxic is called dose-response relationship, and it will be covered in more detail in the following sections.

When we talk about the toxicity of a chemical, we are considering its potential effects on an individual's health, that is, its toxicological hazard. However, the potential of a chemical to cause harm does not mean that this potential will be expressed. The risk to the health depends on the toxicity of a chemical but only appears after exposure to the said chemical by common entry routes to the body: oral, dermal, subdermal (injection), and respiratory route.

As mentioned previously, chemicals that enter the body can cause different adverse effects ranging from more acute such as acute lethality, specific organ toxicity, corrosion or irritation, to chronic such as mutagenetic, teratogenic, carcinogenic, or sensitization effects. Except for the specific organ toxicity where the adverse effect occurs at only one site, both acute and chronic effects are known as system toxicity, i.e., they may occur at multiple sites in the body.

Acute toxicity is the ability of a chemical to cause an adverse effect either after a single exposure or multiple exposures to the substance by any route in a short period of time (<24 hours). In order to assess the acute toxicity of a chemical, it is necessary to generate information on lethality, which was traditionally expressed by measuring levels of exposure ($LC_{50}$) or dose ($LD_{50}$) estimated to kill 50% of a specific population of animals under controlled conditions, and dose-response relationships. Due to ethical and economic reasons, alternative methods for testing acute toxicity such as suites of databases; assays; models; and tools based on modern in vitro, nonmammalian in vivo, and in silico approaches are increasingly being applied.[6]

Corrosion of the living tissues is the acute effect of a chemical that causes substantial irreversible damage to the cells, i.e., cytotoxicity, resulting in the formation of necrotic tissue that can act as a reservoir for the chemical and cause further damage. The chemicals that cause immediate corrosion upon exposure are strong acids/bases, and strong oxidizing/reducing agents.

Certain chemicals can irritate the skin, eyes, or lungs, and cause allergic reactions upon exposure. These effects are usually not as severe as corrosion, and are reversible, resulting in the restoration of the normal function of the affected area or organ. Some examples of acutely toxic chemicals are shown in Fig. 1.2.

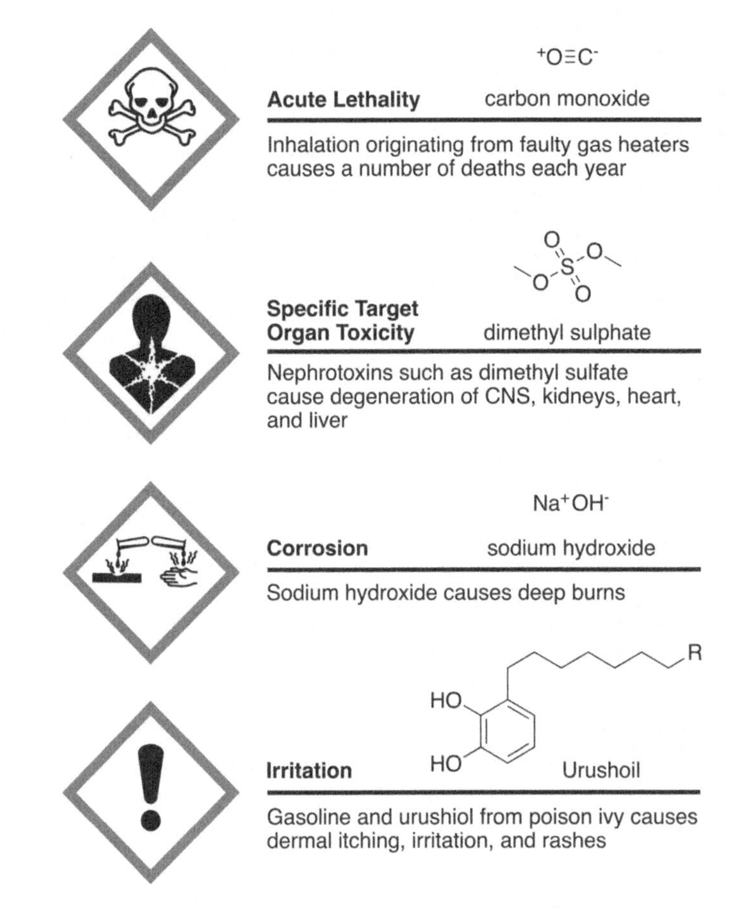


**Figure 1.2** Examples of chemicals causing acute toxicity.[7]

Chronic toxicity is the adverse effect caused by the chemical resulting from repeated, low-level daily doses over a person's or animal's lifetime. Often, damages caused by these effects may go unnoticed for a long time, with repeated exposures slowly building up the damage until it becomes a recognizable clinical disease. Chronic toxicity tests are mainly conducted on test animals or assays over the period of exposure of usually 6 months to 2 years and are often designed to assess carcinogenic potential and the cumulative toxicity of chemicals.

Genetic toxicity causes altered genetic expression by changing or damaging DNA, otherwise known as mutagenesis. Genetic changes are exhibited through either a change in DNA sequence within a gene—gene mutation, changes in chromosome structure—chromosome aberration, or increase/decrease in a number of chromosomes—aneuploidy/polyploidy. If the mutation happens in a somatic cell, it can cause uncontrolled cell growth (cancer) or cell death, and if it happens in a germ cell, the effect can be passed to future generations. Genetic toxicity tests are used to assess the potential of the chemical to induce gene mutations or chromosome damage using bacterial and mammalian in vitro assays, in vivo animal systems, and in silico structure-activity relationship (SAR) models.

Developmental toxicity represents adverse toxic effects of a chemical that affect the developing embryo or fetus. These effects can be expressed either by inducing mutations in the parent's germ cells resulting in abnormal embryos or directly on cells of the embryo leading to cell death or abnormal organ development. The developmental toxicity can result in irreversible conditions leaving permanent birth defects in live offspring, i.e., teratogenicity, embryo growth retardation or delayed growth of specific organ systems, i.e., embryotoxicity, or failure to conceive, spontaneous abortion or stillbirth, i.e., embryolethality.

Carcinogenicity represents a complex process of abnormal cell growth and differentiation that can lead to cancer occurring in two stages—initiation and promotion. Thus, chemicals can act either as initiators, causing the irreversible changes to normal cells, or as promoters, stimulating the initiated cells to progress to cancer growth. The initial transformation of normal cells comes from the mutation of cellular genes that control normal functions which can lead to abnormal cell growth, i.e., to tumors. Tumors can be benign, that do not metastasize and invade other tissues, or malignant (cancer) that can migrate to distant sites which are difficult to treat and can cause death.[8]

Sensitization is a physical process in which a chemical induces the development of an allergic response in an organism exhibited as a rash, difficulty with breathing, or some other reaction. This process is composed of two phases: initial exposure that generates a response to contact with allergens, and the allergic response when the previously sensitized individual is exposed to the allergen again. The initial exposure to a sensitizing chemical might not result in a significant response, however once sensitized, subsequent exposures to even low concentrations can result in severe allergic reactions. Some chemicals are not sensitizers but are irritants that can trigger or enhance the allergic response of sensitized individuals. Most common reactions include asthma, bronchitis, contact rash or eczema, conjunctivitis, and rhinitis. Some examples of acutely toxic chemicals are shown in Fig. 1.3.

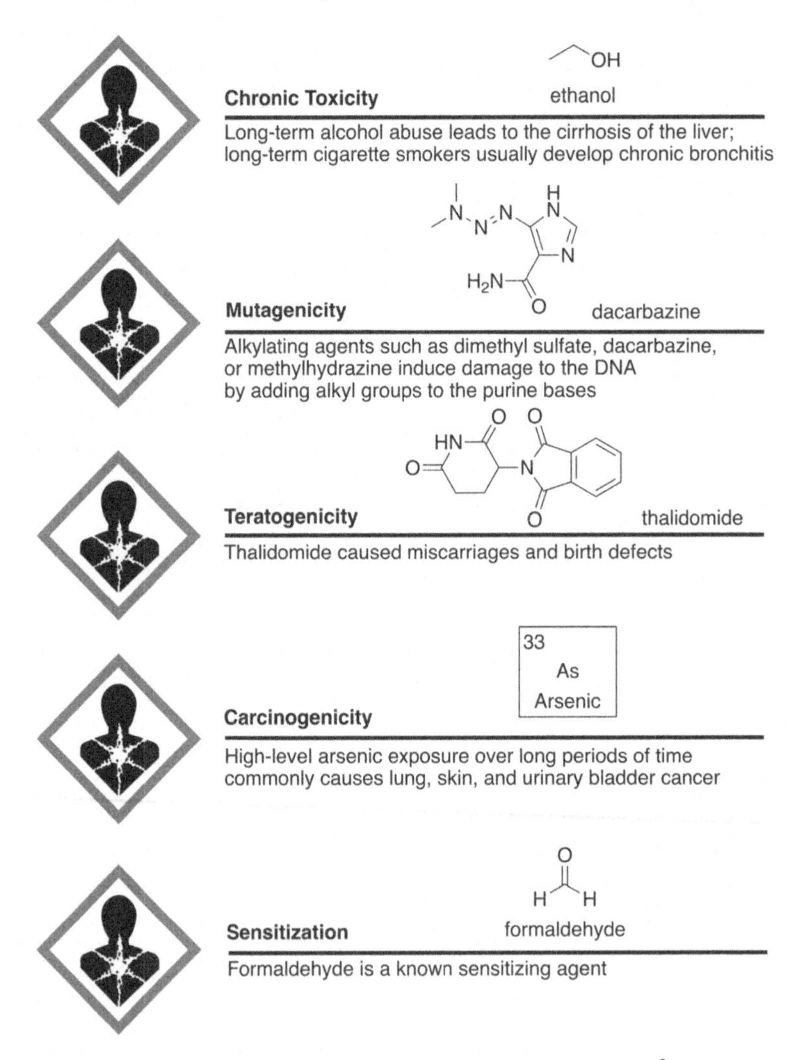

OH
Chronic Toxicity
ethanol
Long-term alcohol abuse leads to the cirrhosis of the liver;
long-term cigarette smokers usually develop chronic bronchitis
Mutagenicity
dacarbazine
Alkylating agents such as dimethyl sulfate, dacarbazine,
or methylhydrazine induce damage to the DNA
by adding alkyl groups to the purine bases
Teratogenicity
thalidomide
Thalidomide caused miscarriages and birth defects
33
As
Arsenic
Carcinogenicity
High-level arsenic exposure over long periods of time
commonly causes lung, skin, and urinary bladder cancer
Sensitization
formaldehyde
Formaldehyde is a known sensitizing agent


**Figure 1.3** Example of chemicals causing chronic adverse effects.[9]

It is worthy of note that large numbers of chemicals exhibit toxic effects in the exposure time of several weeks or months resulting in subchronic toxicity. This pattern is common in human exposure to some environmental agents or pharmaceuticals. For example, workplace exposure to lead for several weeks can lead to anemia.[10]

## 1.3 What Is an Ecotoxicological Hazard?

Once the chemical possessing the potential for causing adverse effects gets released into the environment, it becomes a risk to the organisms living in it, and to the environment itself. Adverse effects causing changes in the state or dynamics of an organism, or at other levels of the biological organization after the exposure to a chemical, are considered ecotoxicological effects and said chemicals—ecotoxicological hazards.[11] The other levels of the organization may include sub-cellular and cellular levels, tissues, individual organisms, populations, communities and ecosystems, and on the largest scale—landscapes (biosphere).[12] At the lowest organizational level, foreign chemicals that enter the body, i.e., xenobiotics, can affect gene transcription by binding to various receptors (such as estrogen receptors), damage proteins and DNA, and promote oxidative stress by increasing the production of reactive oxidative species. At the cellular level, xenobiotics may impair the mitochondrial energy metabolism, accumulate and cause damage by binding to lysosomes, or fragment the chromosomes that are not incorporated in the nucleus.[13] There are a number of xenobiotics that target specific organs or tissues such as the gills of nonmammalian aquatic species causing both acute and chronic adverse effects.[13,14] The ecotoxicological hazards at the organism level rarely cause immediate lethality, instead, most impacts are long-term, affecting susceptibility to diseases, behavior, reproduction, and development. These effects are usually transferred to the upper levels of the organization where they cause changes in population genetics and demographics, affect the communities' structure by altering population interactions, and even cause significant shifts in whole ecosystems dynamics. Figure 1.4 shows two chemicals that are ecotoxicological hazards.[15]

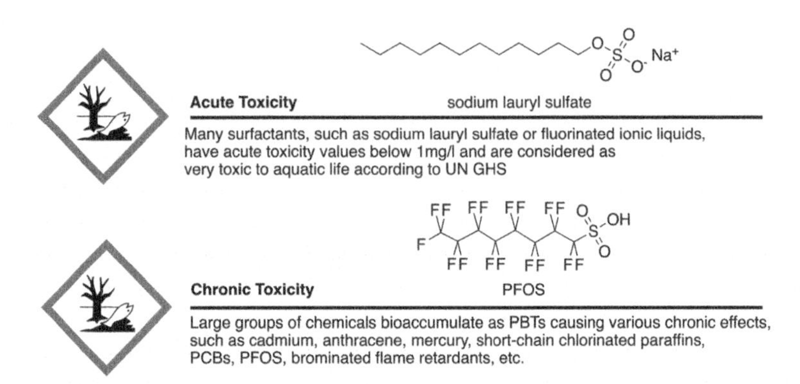


**Figure 1.4** Examples of chemicals that are ecotoxicological hazards.[15]

The adverse effects caused by ecotoxicological hazards can be either acute or chronic. Traditionally, acute toxicity of a chemical was mostly assessed through the investigations on several selected aquatic species (e.g., fathead minnow) as a consequence of the large environmental disasters resulting from the chemical spills into the freshwater systems and other advantages of these organisms for toxicity studies.[16] However, besides aquatic toxicity, exposure to chemicals can lead to toxicity in other environments such as sediment toxicity, terrestrial toxicity, and even respiratory toxicity. The data needed for ecotoxicity assessment can be obtained by various methods such as the use of bioassays, high throughput assays, or by in silico methods such as quantitative structure-activity relationships (QSARs).[17] Aquatic bioassays include water column assays based on *Daphnia*, freshwater and marine fish models, sediment assays based on oyster and *Hyallela* models, and amphibian assays based on frog thyroid models. Terrestrial assays include standardized studies for various types of birds, hard- and soft-bodied soil invertebrates, bees, and germination and growth of plants. In the case of terrestrial mammals, hazards are determined through the data generated for human health assessments mostly using rodent models. In the literature, most of the ecotoxicity data exist for the acute effects on aquatic freshwater species, with increasing numbers of toxicity testing on plants and soil invertebrates. On the other hand, toxicity data for terrestrial vertebrates is pretty sparse in part due to ethical and economic reasons.

## 1.4 What Is a Physical Hazard?

Chemicals that due to their intrinsic properties cause physical harm or damage are considered physical hazards. This hazard group contains a broad range of chemicals that are flammable, self-heating, self-reactive, explosive, corrosive to metals, or act as oxidizers (Fig. 1.5).[18]

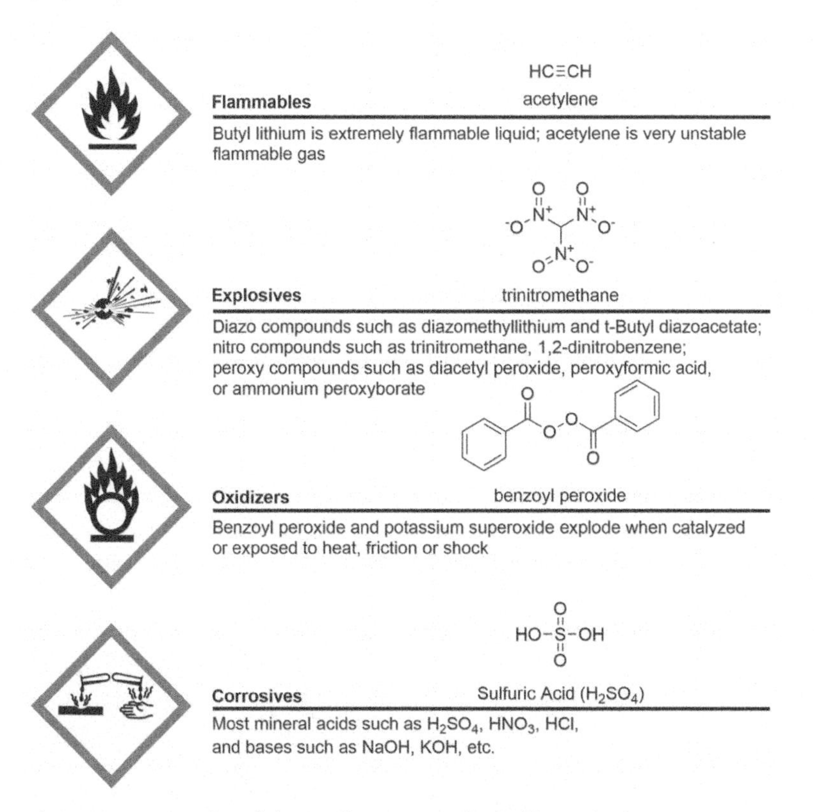

HC≡CH
Flammables
acetylene
Butyl lithium is extremely flammable liquid; acetylene is very unstable flammable gas
Explosives
trinitromethane
Diazo compounds such as diazomethyllithium and t-Butyl diazoacetate; nitro compounds such as trinitromethane, 1,2-dinitrobenzene; peroxy compounds such as diacetyl peroxide, peroxyformic acid, or ammonium peroxyborate
Oxidizers
benzoyl peroxide
Benzoyl peroxide and potassium superoxide explode when catalyzed or exposed to heat, friction or shock
HO-S-OH
Corrosives
Sulfuric Acid ($H_2SO_4$)
Most mineral acids such as $H_2SO_4$, $HNO_3$, HCl, and bases such as NaOH, KOH, etc.


**Figure 1.5** Examples of chemicals acting as physical hazards.

Flammable chemicals pose a risk because they can ignite easily, generate a large amount of heat, and enable the easy spread of fire. Additionally, they can act as health hazards since some thermal degradation or combustion products are toxic or respiratory irritants, and flame causes oxygen to be consumed. Flammable chemicals can be gases, liquids, and solids. While gases easily burn, liquids and solids first vaporize or thermally degrade and release flammable gases and vapors which in turn burn. In order for the flame to propagate, three components are necessary: an ignition source of minimum temperature, duration, and energy; fuel—gas or vapor of certain concentration; and an oxygen supply. Combustion is initiated when the finite amount of energy (heat) is introduced chemically (by chemical reaction or spontaneous combustion), mechanically (by friction), electrically (by spark or arc), or via radiation (e.g., solar). Some chemicals are very reactive, and they will spontaneously ignite when in contact with air, either under ambient conditions or at elevated temperatures (up to 54 °C for pyrophoric gases), due to rapid oxidation by oxygen or moisture in the air.[1] These chemicals are called pyrophoric, and many of them can also react when they come into contact with water releasing flammable gases. In some cases, usually, when they are confined in some container, flammable chemicals can explode upon ignition.[1]

A lot of chemical reactions are exothermic, i.e., they release energy. This release is gradual when the reaction proceeds slowly, with most of the energy usually released in the form of heat. However, if the reaction is very rapid, a large amount of energy can be released in a short amount of time. This can manifest itself by a rapid expansion of gases at such pressure, temperature, and speed to cause damage to the surroundings, i.e., as an explosion. Chemical explosions can be distinguished from other highly exothermic processes by the extreme rapidity of their reactions.

Explosive chemicals are either liquids or solids in themselves capable of causing explosions, however, this is not a spontaneous process. Usually, they need some amount of input energy to detonate.

Another group of very reactive chemicals are oxidizers. These chemicals, while by themselves are not necessarily combustible, may cause or contribute to the combustion of other substances or materials. They can be gases, liquids, and solids, and generally promote combustion by easily yielding oxygen that is part of their chemical structure. Considering that they are good electron acceptors, they will react promptly in contact with reducing reagents. Depending on their reactivity, oxidizers can be classified as relatively stable, which increase the burning rate of combustible materials; moderately reactive, which cause spontaneous heating of combustible materials, vigorously decompose when heated, and explode if heated in a sealed container; and unstable, which release oxygen at room temperature and explode when catalyzed or exposed to heat, shock, or friction.

Finally, a group of chemicals causing irreversible damage, or even destruction to materials by coming into direct contact, are considered corrosives. A higher concentration of a corrosive chemical usually results in more intensive damage. Corrosives can be gases, liquids, or solids and most of them are (or can act as) strong acids and bases. Due to the large variety of chemicals in this group, many of them can be hazardous in more than one way, additionally being flammable, highly reactive, or causing health issues. Interestingly, some corrosives are commonly used as cleaning agents, and because they tend to be highly reactive, they are often used in a large number of chemical reactions.[19]

## 1.5 What Causes Explosivity?

As hinted in the previous section, while the oxidation reactions that will result in explosions are energetically possible, this process is not spontaneous. Usually, a small energetic barrier needs to be overcome to start the reaction, which will then continue by itself until completed. The energy input necessary to pass this barrier is known as initiation or detonation. Safety regulations often separate explosives by the amount of energy needed for initiation into three categories: primary (sensitive) explosives, that only need a small amount of energy; intermediate explosives, requiring slightly more energy; and secondary (insensitive) explosives, that need relatively more energy to detonate.[20]

Certain types of functional groups are known to be present in the chemical structure of explosive chemicals.[21] These groups usually contain oxygen or nitrogen atoms that are violently released in gas creating a rapidly expanding shock wave. Upon detonation, exothermic reactions convert much of the potential energy stored in chemical bonds into kinetic energy in the form of gas release by combining oxygen atoms with carbon or hydrogen atoms forming gaseous $CO$, $CO_2$, or $H_2O$, while nitrogen atoms are transformed into $N_2$ gas. These reactions are similar to combustion, however, unlike flammables which have an almost unlimited supply of oxygen from the air, explosive chemicals must contain a limited amount of oxidizer within their molecular structure. Due to this fact, detonation releases stored chemical energy much more rapidly compared to combustion.

Some of the functional groups that are considered to potentially contribute to the explosiveness of a chemical compound are shown in Fig. 1.6. These are acetylene (–C=C–), azo (–N=N–), diazo (=N=N), organic/metal azide (R/M-$N_3$), diazonium salts (R–$N_2^+$), fulminate (–C=N–O), halogen amine (=N–X), nitrate (–$ONO_2$), nitro (–$NO_2$), aromatic or aliphatic nitroamine (=N–$NO_2$ or –NH–$NO_2$), nitrite (–ONO), nitroso (–NO), ozonide (–$O_3$–), peracids (–CO–O–O–H), peroxide (–O–O–), hydroperoxide (–O–O–H), metal peroxide (M–O–O–M), nitrogen metal salts (=N–M). Additionally, a number of halogen-oxysalts such as bromate ($BrO_3^-$), chlorate ($ClO_3^-$), chlorite ($ClO_2^-$), perchlorate ($ClO_4^-$), iodate ($IO_3^-$), are potent contributors to explosivity. Thus, it is especially important to avoid these functional groups when designing new chemicals if possible.

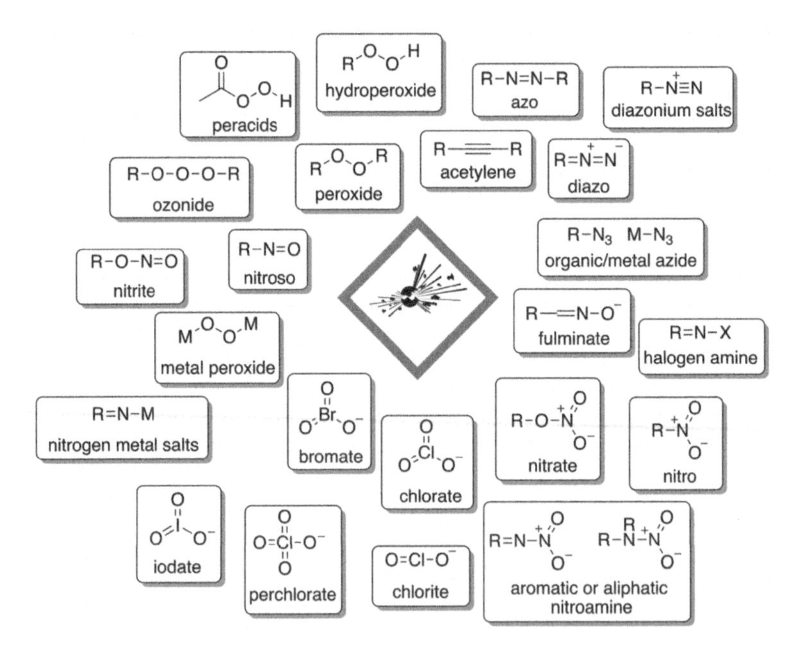


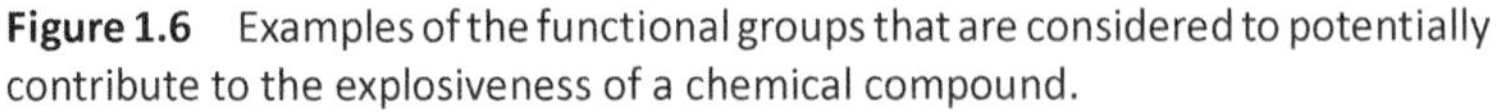

**Figure 1.6** Examples of the functional groups that are considered to potentially contribute to the explosiveness of a chemical compound.

## 1.6 What Is "Oxygen Balance" and How Does It Calculate Explosivity?

Oxygen balance (OB) represents a method to quantify the amount of oxygen present in an explosive and by extension, estimate the degree to which an explosive can be oxidized.[20,21] OB can be defined as a weight% of oxygen liberated as a result of the complete conversion of carbon to CO and $CO_2$, hydrogen to water, sulfur and phosphorus to their respective oxides, nitrogen and halogens to their non-oxidizable products, and all metals to metal oxides. When discussing OB, three forms are considered:

- negative OB—where the explosive molecule contains less oxygen than necessary for the full conversions and a large amount of toxic gases such as carbon monoxide released
- zero OB—where the explosive molecule contains just enough oxygen for the full conversions without excess
- positive OB—where the explosive molecule contains more oxygen than is needed for the full oxidation of its components

OB is mostly computed following the formula [1]:

$$\text{OB\%} = \frac{-1600\left(2x + \dfrac{y}{2} - z\right)}{\text{molecular mass of compound}} \quad (1)$$

where $x$ is the number of carbon atoms in the molecule; $y$ is the number of hydrogen atoms in the molecule, and $z$ is the number of oxygen atoms in the molecule.

Table 1.1 shows examples of chemicals having negative, zero, and positive OB level with three examples showing the calculated OB level using the above formula.

Organic substances containing chemical groups associated with explosive properties are considered explosives if their exothermic decomposition energy is less than 500 J/g or the onset of exothermic decomposition is below 500 °C.

**Table 1.1** Examples of explosive chemicals with regard to OB level[20]

| | | | |
|---|---|---|---|
| Negative OB | Diazomethane, ethyl hydroperoxide, trinitrotoluene | 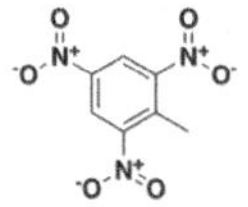 [structure] | trinitrotoluene<br>MW: 227.13 g/mol<br>**OB: -73.97** |
| Zero OB | Trinitrotriazine (a theoretical molecule) | 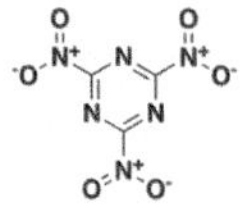 [structure] | trinitrotriazine<br>MW: 216.07 g/mol<br>**OB: 0** |
| Positive OB | Nitroglycerin, ammonium nitrate, lithium perchlorate | 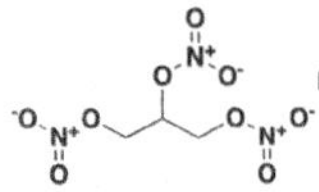 [structure] | nitroglycerin<br>MW: 227.09 g/mol<br>**OB: 3.52** |

## 1.7 What Is a Global Hazard?

Earlier, we described the environmental hazards from the perspective of ecotoxicological impacts. On one hand, once chemicals get into the environment, they can cause a number of adverse effects to various living organisms and have a lasting impact even at the scale of whole ecosystems. On the other hand, these chemicals, whether they are anthropogenic or naturally released, can have a significant impact on a much larger scale, i.e., they can affect the whole planet. The World Health Organization (WHO) defines large-scale and global environmental hazards as those affecting human health including climate change, stratospheric ozone depletion, changes in ecosystems due to the loss of biodiversity, changes in hydrological systems and the supplies of fresh water, land degradation, urbanization, and stresses on food-producing systems.[22] Centuries of human activity, especially since the Industrial Revolution in the 18$^{th}$ century, have contributed to the appearance, or intensification, of these environmental hazards (Fig. 1.7). As a consequence of this activity, large quantities of various hazardous chemicals have found their way into the environment.

One of the major environmental concerns recognized today is rapid climate change and global warming. The rapid increase of global average temperatures of Earth's surface, atmosphere, and oceans was confirmed by thousands of studies in recent decades. There is ample evidence that human activities, like the emissions of greenhouse gases from fossil fuel combustion, are the primary driver of the observed climate changes. The concentration of $CO_2$ in the atmosphere is larger than any time in the last 800,000 years, and since the Industrial Revolution began, it is estimated that the $CO_2$ level increase was more than 30 % (Fig. 1.8).[24] The impacts of these changes can already be felt all around the world and these impacts are only expected to increase. The rising temperatures that are affecting whole biomes, rises in sea levels, and the acidifying of the world oceans are only some of the consequences of rapid climate change.

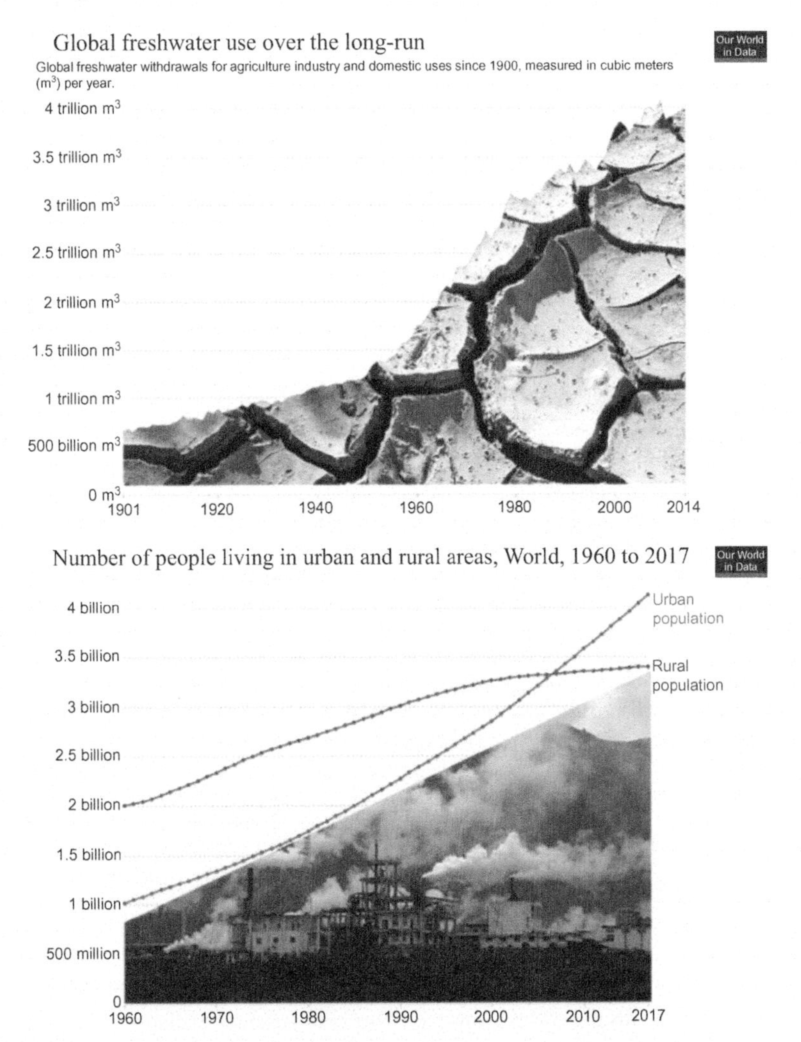


**Figure 1.7** Intensification and concentration of human activities.[23]

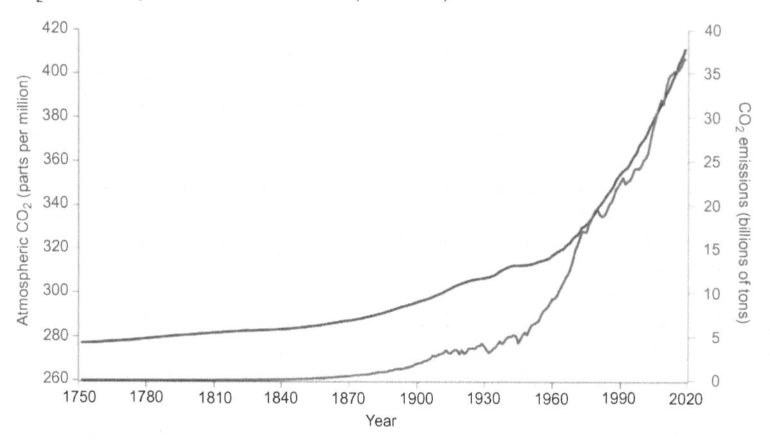


**Figure 1.8** Human-caused climate change: The amount of carbon dioxide in the atmosphere (red line) and human caused emissions (blue line). Adapted from the original by Dr. Howard Diamond (NOAA ARL).[24]

Another illustrative example, that was in part mitigated by the mobilization of the scientific community and combined efforts of world governments, is the appearance of the ozone hole. The ozone layer, located in the stratosphere, has protected life on Earth for millions of years from the harmful effects of UV-B and UV-C radiation. In the 1980s, scientists[25] observed the reduction of concentrations of ozone above the Antarctic. The ozone "hole" steadily increased in size and length of existence, until the United Nations Montreal Protocol[26] limited production and consumption of ozone-depleting compounds, chlorofluorocarbons, carbon tetrachloride, halons, and methyl chloroform. Most of these compounds were used in refrigeration, due to their chemical stability and low boiling points. However, this stability enabled them to reach the stratosphere where they were split by intense UV radiation, losing a chlorine atom that in turn reduced the ozone to oxygen. Additionally, thousands of ozone molecules could be broken down by a single chlorine atom acting as a replenishable catalyst.[27]

Another important environmental hazard that we are encountering in modern society is the depletion of various resources (Fig. 1.9). As technology advances and the human population increases, the need for more resources is rising steeply as well. While it might seem that the raw resources used to satisfy the needs of various industries (agricultural, chemical, electronic, etc.) are infinite, it has increasingly become evident that a lot of these resources are getting much harder to obtain, causing more environmental damage in the extraction and refinement process. Some obvious examples are the diminishing fossil fuel reserves or the limited access to rare earth elements used in the electronic industry. That said, it is important to emphasize that efforts to find sustainable and environmentally friendly alternatives to limited resources are a crucial part of the green chemistry mission and an integral part of initiatives supported by numerous government, educational, and research institutions and industries.

**Figure 1.9** Large-scale resource extraction across the world.[28]

## References

1. United Nations. *Globally Harmonized System of Classification and Labelling of Chemicals (GHS)*; ST/SG/AC.10/30/Rev.8; United Nations: New York and Geneva, 2019, p. 570.
2. United States Occupational Safety and Health Administration. Chemical hazards and toxic substances, *Hazard Communication*, Std. No. 1910.1200, United States Department of Labor: Washington, DC, 2012.
3. UNECE. Globally harmonized system of classification and labeling of chemicals, GHS, Rev. 8, United Nations, 2019.
4. Ohrel, Jr., R. L.; Register, K. M. Chapter 12: Contaminants and toxic chemicals heavy metals, pesticides, PCBs, and PAHs. *Volunteer Estuary Monitoring Manual: A Methods Manual*, 2nd ed., EPA-842-B-06-003, 2006.
5. Winter, C. K.; Francis, F. J. Assessing, managing, and communicating chemical food risks. *Food Technol.*, 1997, **51**(5), 85.
6. National Academies of Sciences, Engineering and Medicine. *Application of Modern Toxicology Approaches for Predicting Acute Toxicity for Chemical Defense*. The National Academies Press: Washington, D.C., 2015, p. 135.
7. (a) Byard, R. W. Carbon monoxide: the silent killer. *Forensic Sci Med Pathol*, 2019, **15**(1), 1–2; (b) Vyskocil, A.; Viau, C. Dimethyl sulfate: review of toxicity. *CEJOEM*, 1999, **5**(1), 72–82.
8. Weston, A.; Harris, C. C. Multistage carcinogenesis. In: Kufe, D. W.; Pollock, R. E.; Weichselbaum, R. R.; Bast, Jr., R. C.; et al. (Eds.), *Holland-Frei Cancer Medicine*, 6th ed., BC Decker: Hamilton, ON, 2003.
9. Klaassen, C. D.; Watkins, III, J. B. *Casarett & Doull's Essentials of Toxicology*, 3rd ed., McGraw-Hill Education: New York, 2015, p. 524.
10. Staudinger, K. C.; Roth, V. S. Occupational lead poisoning. *Am Fam Physician*, 1998, **57**(4), 719–726.
11. van Leeuwen, C. J. Ecotoxicological effects. In: van Leeuwen, C. J.; Hermens, J. L. M. (Eds.), *Risk Assessment of Chemicals: An Introduction*, 1st ed., Springer: Dordrecht, 1995, p. 374.
12. National Research Council. *A Framework to Guide Selection of Chemical Alternatives*. The National Academies Press: Washington, DC, 2014.
13. Hutchinson, T. H.; Madden, J. C.; Naidoo, V.; Walker, C. H. Comparative metabolism as a key driver of wildlife species sensitivity to human and

veterinary pharmaceuticals. *Philos Trans R Soc Lond B Biol Sci*, 2014, **369**(1656), 20130583.

14. Walker, C. H.; Sibly, R.; Hopkin, S. P.; Peakall, D. B. *Principles of Ecotoxicology*. CRC Press, 2012.

15. (a) Kenedy, G. L. Surfactants, anionic and nonionic. In: Wexler, P. (Ed.), *Encyclopedia of Toxicology*, 3rd ed., Elsevier: Amsterdam, 2014, pp. 436–438; (b) Vieira, N. S. M.; Stolte, S.; Araujo, J. M. M.; Rebelo, L. P. N.; et al. Acute aquatic toxicity and biodegradability of fluorinated ionic liquids. *ACS Sustain Chem Eng*, 2019, **7**(4), 3733–3741.

16. Ankley, G. T.; Villeneuve, D. L. The fathead minnow in aquatic toxicology: past, present and future. *Aquat Toxicol*, 2006, **78**(1), 91–102.

17. (a) Voutchkova, A. M.; Kostal, J.; Steinfeld, J. B.; Emerson, J. W.; et al. Towards rational molecular design: derivation of property guidelines for reduced acute aquatic toxicity. *Green Chemistry*, 2011, **13**(9), 2373–2379; (b) Voutchkova, A. M.; Osimitz, T. G.; Anastas, P. T.; Toward a comprehensive molecular design framework for reduced hazard. *Chem Rev*, 2010, **110**(10), 5845–5882; (c) Voutchkova-Kostal, A. M.; Kostal, J.; Connors, K. A.; Brooks, B. W.; et al. Towards rational molecular design for reduced chronic aquatic toxicity. *Green Chem*, 2012, **14**(4), 1001–1008.

18. Wilrich, C.; Brandes, E.; Michael-Schulz, H.; Schröder, V.; et al. UN-GHS—physical hazard classifications of chemicals: a critical review of combinations of hazard classes. *J Chem Health Saf*, 2017, **24**(6), 15–28.

19. Gerster, F. M.; Vernez, D.; Wild, P. P.; Hopf, N. B. Hazardous substances in frequently used professional cleaning products. *Int J Occup Environ Health, 2014,* **20**(1), 46–60.

20. Oxley, J. C. The chemistry of explosives. In: Zukas, J. A.; Walters, W. P. (Eds.), *Explosive Effects and Applications*, Springer: New York, NY, 1998, pp. 137–172.

21. Lothrop, W. C.; Handrick, G. R. The relationship between performance and constitution of pure organic explosive compounds. *Chem Rev*, 1949, **44**(3), 419–445.

22. WHO. Global environmental change. https://www.who.int/globalchange/environment/en/ (accessed April 30).

23. (a) Contrast, H. A factory in China at Yangtze River (changes made, CC BY 2.0 DE). https://commons.wikimedia.org/wiki/File:Factory_in_China.jpg (accessed April 30); (b) Kaufmann, B. Drought (changes made, CC BY 2.0). https://www.flickr.com/photos/22746515@N02/3487433937 (accessed April 30); (c) Ritchie, H.; Roser, M. Water

use and stress (changes made, CC BY 4.0). https://ourworldindata.org/water-use-stress (accessed December 1); (d) Ritchie, H.; Roser, M. Urbanization (changes made, CC BY 4.0). https://ourworldindata.org/urbanization (accessed December 1).

24. NOAA Climate.gov; https://www.climate.gov/news-features/understanding-climate/climate-change-atmospheric-carbon-dioxide (accessed April 30).
25. Farman, J. C.; Gardiner, B. G.; Shanklin, J. D. Large losses of total ozone in Antarctica reveal seasonal $ClO_x/NO_x$ interaction. *Nature*, 1985, **315**(6016), 207–210.
26. United Nations Environment Programme. About Montreal Protocol. https://www.unenvironment.org/ozonaction/who-we-are/about-montreal-protocol (accessed September 1, 2020).
27. Dessler, A. *Chemistry and Physics of Stratospheric Ozone*. Elsevier Science & Technology, 2000.
28. (a) Hodge, K. Oil rig maintenance in Darwin harbour, May 2006 (change made, CC BY 2.0). https://www.flickr.com/photos/40132991@N07/4420332059 (accessed April 30); (b) crustmania. Deforestation (changes made, CC BY 2.0). https://www.flickr.com/photos/76771610@N00/233523196 (accessed April 30); (c) Kaufmaan, B. Open coal mine Garzweiler II (changes made, CC BY 2.0). https://www.flickr.com/photos/22746515@N02/3281728778 (accessed April 30).

# Chapter 2

# ADME

## 2.1 Why Is Absorption Crucial to Toxicity?

Chemical compounds from the environment, i.e., xenobiotics, can get into the body upon exposure and then express adverse effects—toxicity. These compounds can interact with specific molecules in the cell and cause cellular dysfunction or change the biological environment that will result in molecular, cellular, or organ dysfunction. Large numbers of chemicals penetrate the epithelial barriers via diffusion, but they can also be transported into the body upon ingestion via the gastrointestinal tract. Whether a particular molecule will be absorbed (and in what quantity) is influenced by several factors like the concentration of the substance, its physicochemical properties, and the route of exposure.[1]

The main routes for the absorption of xenobiotics into the body are respiratory, dermal, and gastrointestinal, but others exist, such as subdermal injections or by implantation (Fig. 2.1). Table 2.1 show some typical examples of chemicals that get absorbed through these main routes of exposure.

*First Do No Harm: A Chemist's Guide to Molecular Design for Reduced Hazard*
Predrag V. Petrovic and Paul T. Anastas

ISBN 978-981-4968-59-1 (Hardcover), 978-1-003-35964-7 (eBook)
www.jennystanford.com

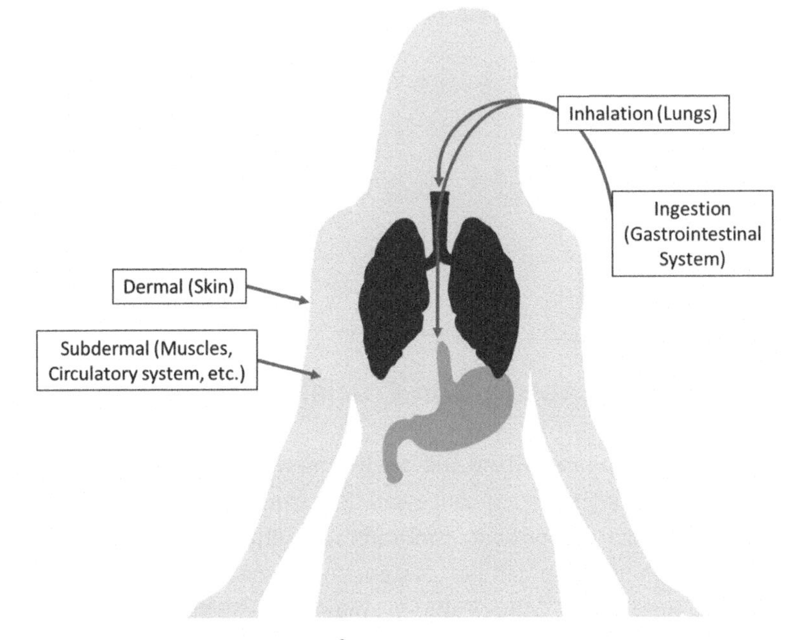


**Figure 2.1** Routes of exposure.[2]

**Table 2.1** Examples of chemicals easily absorbed through main routes of exposure[1]

| | |
|---|---|
| Lungs | Air pollutants such as carbon monoxide, ozone, hydrocarbons, sulfur oxide get easily absorbed by inhalation |
| Skin | Dimethyl sulfoxide (DMSO) is quickly absorbed through the skin; semi-volatile organic compounds such as chlorpyrifos, diethyl phthalate or butylated hydroxytoluene are often found in everyday household items increasing the risk for exposure as they are easily absorbed through the skin |
| GI tract | Weak organic acids are easily absorbed in the stomach; thallium, paraquat and lead are actively transported through the intestinal wall and distributed throughout the body |

The hazard level of a chemical is often defined by the route of exposure since, depending on the concentration and properties of the substance, some routes can lead to high toxicity while others will not result in high toxicity. One example is the pesticide dichlorodiphenyltrichloroethane (DDT), which when exposed to the skin in a powdered form, will not get absorbed in as large of a quantity as when ingested.[3]

The respiratory route is the path by which chemicals which are gases, vapors of volatile liquids, or aerosols are absorbed into the body via the lungs. However, gases or vapors can be absorbed in the mucosa of the nose if they react with the cells or are very water-soluble. When these xenobiotics get into the lungs, they diffuse into the blood through the alveoli until equilibrium is reached. Once in the blood, the circulatory system will distribute any absorbed chemical through the body to different organs where they can exhibit toxicity. The fate of aerosols and particle depends on their size. Most larger particles will be filtered in the nose, those approximately 2.5 μm in diameter get deposited in the mucus layer of the bronchia and trachea where they can get removed or swallowed and absorbed in the GI tract, 1 μm and smaller can penetrate the alveoli, while nanoparticles 10–20 nm have the highest potential to get absorbed into the blood (Figure 2.2).[4]

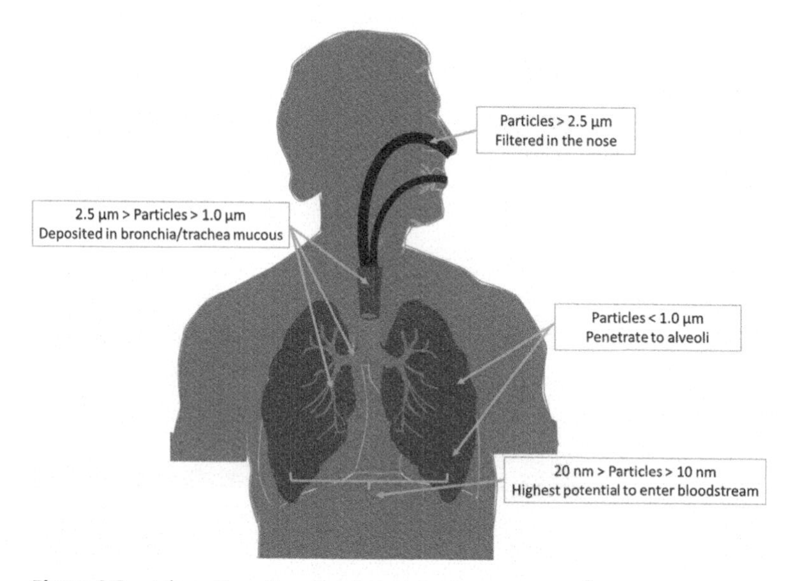


**Figure 2.2** Adsorption sites via inhalation exposure route.[5]

The skin is a very important exposure route since many pharmaceutical and consumer products are dermally applied. It is the human body's largest organ, and it is the first line of defense against all outside hazards. Before the xenobiotic can enter the body and cause adverse effects on the organs and systems inside, it needs to traverse the several layers comprising the skin tissue. This tissue is relatively resistant to the permeation of aqueous solutions and most ions, thus being a barrier to most xenobiotics. The skin consists of three main layers of cells: epidermis, dermis, and hypodermis. The epidermis is the outermost layer responsible for the regulation of contaminant penetration. This is achieved by the presence of several layers of the keratin-packed cells that are chemically resistant material. The thickness of this layer varies across the body, affecting the ability of xenobiotics to penetrate and get to the blood vessels below the epidermis. Thus, any process that damages or destroys the top layer of skin will make it easier for any toxicant to enter the body. Chemicals move through the epidermis by passive diffusion, where polar compounds diffuse through the hydrated keratinized layer and nonpolar through the lipid material between the keratin filaments (Fig. 2.3).[6] Small amounts of chemicals (mostly hydrophilic) can get to the bloodstream through glands and hair follicles, containing pores going through the epidermis to the dermis layer. Once chemicals get to the dermis layer, they can easily diffuse to the blood vessels and get into the systemic circulation. Factors that influence the ability of a chemical to be absorbed through the skin are the size of the molecule, ambient temperature, presence of the carrier solvents, hydration state of the keratin layer, and the integrity of the epidermis.[4]

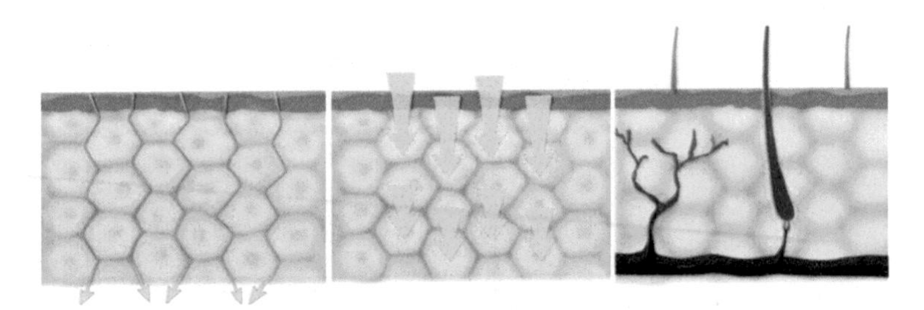

**Figure 2.3** Intercellular, transcellular, and appendage permeation pathways.[6]

A large number of chemicals from the environment get into the body by ingestion through the contaminated food or drinks and get absorbed by the gastrointestinal (GI) tract. The GI tract consists of several areas: mouth and esophagus, stomach, intestine, colon, and rectum (Fig. 2.4). The main factors that influence the

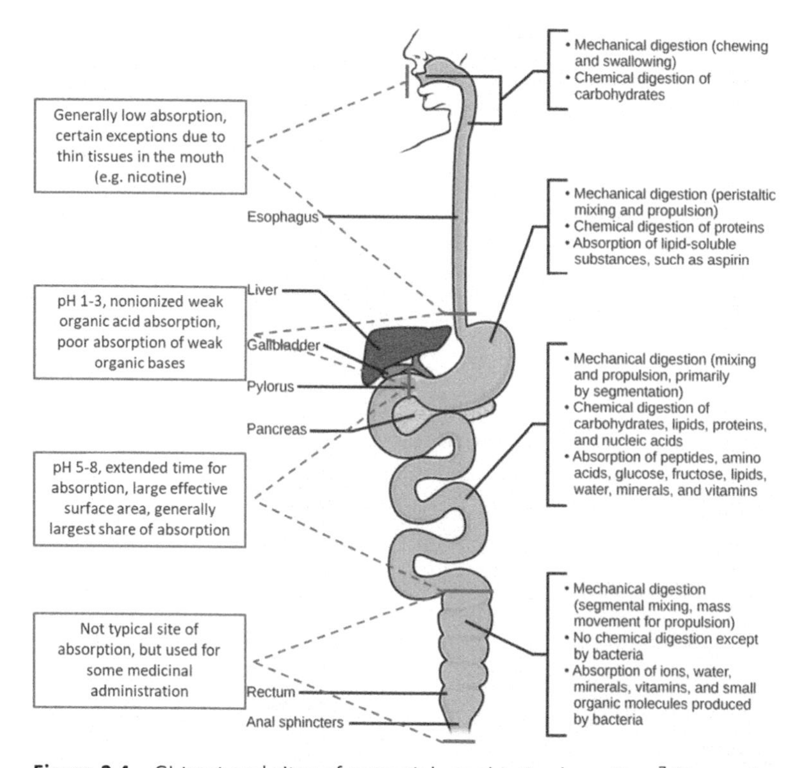


**Figure 2.4** GI tract and sites of potential xenobiotic absorption.[7] (Access for free at https://openstax.org/books/biology-2e/pages/1-introduction)

absorption through the GI tract are the type of cells, pH of stomach or intestine area, and period of time that the substance remains at the site of absorption. The absorption of xenobiotics in the mouth and esophagus is generally low mostly due to the short time they spend in these areas. However, some tissues in the mouth are very thin, allowing for the absorption of certain chemicals (e.g., nicotine, nitroglycerine). The high acidity of the stomach (pH 1–3) makes it a good site for absorption of nonionized weak organic acids and a bad site for the absorption of highly ionized weak organic bases. Most of the absorption by the GI tract takes place in the intestine. This organ has evolved with the role of absorbing nutrients, with a very large effective surface area containing a thin layer of cells in mucosa connected by a large number of blood vessels. A pH of 5–8 makes the passive diffusion for both nonionized weak acids and bases, and for the small lipid-soluble molecules, very efficient. Many essential nutrients, but also toxicants, are moved to the body by active transport, and additionally, the slow movement of materials through the intestine contributes to the prolonged exposure and increased length of time for absorption. Xenobiotics are usually not absorbed in the colon and rectum, but this site is sometimes used for the administration of some medicines.

## 2.2 What Is Lipinski's Rule of Five?

The original rule of five was devised to assess the "drug-like" character for orally active compounds based on simple physicochemical parameter ranges that are associated with the acceptable intestinal permeability and aqueous solubility to be used by the medicinal chemists.[8] The rule was created by analyzing thousands of phase II drug compounds available at the time and the distribution of calculated properties among these compounds and considered as heuristic, i.e., rule-of-thumb approach.

The rule states that the molecule will more likely exhibit poor permeation or absorption if its molecular mass is over 500 Daltons; the calculated log P is over 5, i.e., they exhibit high lipophilicity; there are more than 5 hydrogen bond donors counted as a sum of OH and NH groups in the molecular structure, and there are more than 10 hydrogen bond acceptors counted as a sum of N and O atoms in the molecular structure.[8]

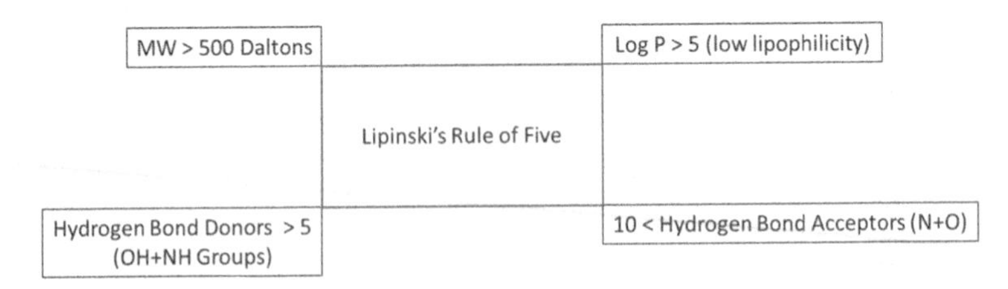


**Figure 2.5** Lipinski's rule of five.

These parameters (Fig. 2.5) were intentionally created as a conservative predictor of bioavailability to guide the chemical design away from the medicinal and combinatorial chemistry approach of the time that resulted in a large number of chemical compounds with poor physicochemical properties.[9] Failing the rule of five test can indicate an oral activity problem for a compound, however, passing the test does not mean automatic "drug-likeness". Additionally, this rule does not address structural features that cause specific chemical behavior of drug and non-drug compounds. This resulted in some existing orally active drugs such as antibiotics, antifungals, cardiac glucosides, and vitamins violating the rule because these compounds have structural features that make them a substrate for naturally occurring transporters. The rule does not estimate oral absorption in a quantitative matter but provides simple and quick guidance for the drug candidate screenings.[10] Over the years the rule of five was expanded to improve the predictions by including various extensions such as those introduced by Ghose et al.,[11] Veber et al.,[12] or the rule of three.[13]

## 2.3 What Is Log P/Log $K_{ow}$?

As we mentioned earlier, in order for a chemical to enter the body and express its adverse effects, it needs to pass through the various biological membranes of a number of cells. The majority of a cell membrane consists of a lipid bilayer formed by phospholipids, glycolipids, and cholesterol. Phospholipids are amphiphilic molecules with hydrophilic polar heads, oriented towards inside and outside of the cell, and hydrophobic lipid tail, interacting with other fatty acids forming the continuous hydrophobic space of the cell membrane. Through the membrane, numerous proteins are embedded or inserted to fulfill the role of receptors, transporters, aqueous pores, or ion channels.[7] A schematic representation of the cell membrane is shown in Fig. 2.6.

Once the chemical contacts the cell, it can get inside either by passive diffusion or by active transport. Small hydrophilic molecules diffuse through aqueous pores, while larger hydrophobic molecules diffuse through the lipid bilayer. The rate of transport across the membrane for hydrophobic molecules correlates with their lipophilicity, i.e., lipid solubility.[14] To assess this solubility, first, it is necessary to define the ratio of concentrations for a given compound in a mixture of two immiscible solvents at equilibrium, i.e., partition coefficient P.[14] Considering that the cell membrane is essentially a boundary between organic and aqueous phases, partition coefficient

P is usually defined as a ratio of the concentration of a nonionic compound in n-octanol and water at equilibrium at a specified temperature, i.e., $K_{ow} = C_{in\ octanol}/C_{in\ water}$. The $K_{ow}$ partitioning coefficient is commonly expressed in the logarithmic form to base of 10 and indicated as log $K_{ow}$ or log P.[15] This parameter is also widely used to assess potential membrane permeability. It can be obtained using experimental procedures (liquid–liquid extraction or HPLC), estimation techniques, or estimated by in silico methods (based on the structure of the molecule).

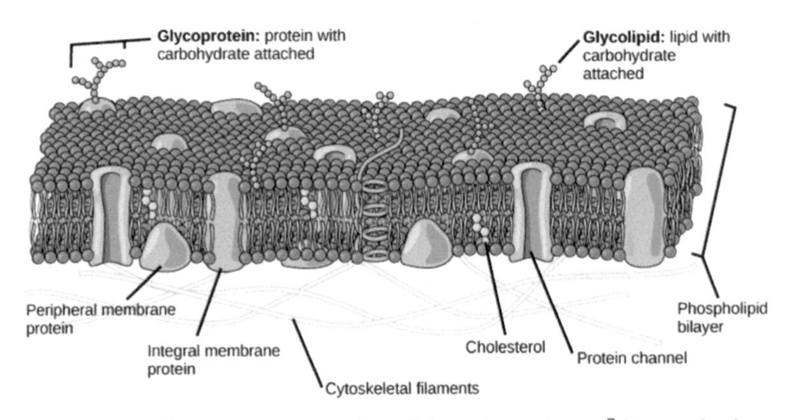


**Figure 2.6** Schematic representation of the cell membrane.[7] (Access for free at https://openstax.org/books/biology-2e/pages/1-introduction)

## 2.4 How Does Log P Affect Absorption?

As lipophilicity has become an important descriptor in understanding the transport and impact of chemicals in ecological and physiological systems, measurement or prediction of log P values found their way to many areas of research and industries. A positive value of log P indicates a higher concentration of a chemical in a lipid phase, log P = 0 means that the chemical is equally partitioned between the lipid and aqueous phases, while negative log P indicates a higher affinity for the aqueous phase. Considering the absorption, log P values provide important information on the bioavailability of chemicals. We already mentioned that, for oral drugs, Lipinski's rule of five states that the chemicals with log P values larger than

5 usually exhibit poor absorption. On the cellular level, smaller log P values indicate the ability of a chemical to penetrate the cell wall, i.e., lipid bilayer, and get transported across the various biological membranes. This property is very important for all the drug candidates, as they need to be lipophilic enough to penetrate the cell wall (log P = 2–4), but not so much to get embedded in it (log P > 5–6). Examples of log P values for several common pharmaceutical are shown in Figure 2.7. Of course, this also means that any other xenobiotic with appropriate lipophilic character will behave in the same way, and potentially cause adverse effects.

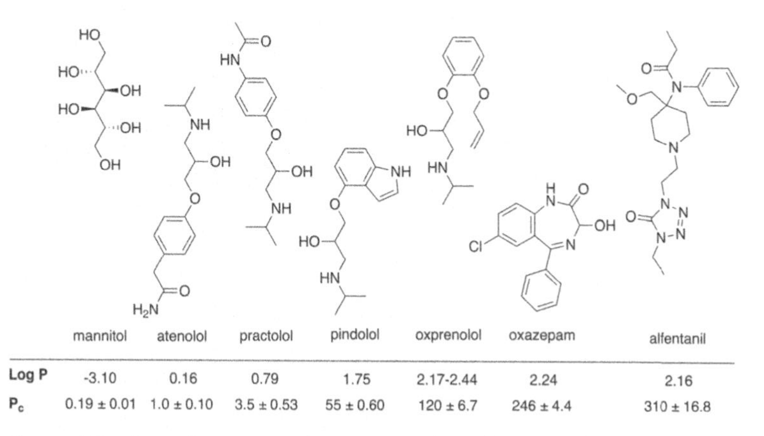

| | mannitol | atenolol | practolol | pindolol | oxprenolol | oxazepam | alfentanil |
|---|---|---|---|---|---|---|---|
| **Log P** | -3.10 | 0.16 | 0.79 | 1.75 | 2.17-2.44 | 2.24 | 2.16 |
| $P_c$ | 0.19 ± 0.01 | 1.0 ± 0.10 | 3.5 ± 0.53 | 55 ± 0.60 | 120 ± 6.7 | 246 ± 4.4 | 310 ± 16.8 |

**Figure 2.7** Several common pharmaceuticals with varying log P values and absorption rates. Adapted with permission from Ref. [16]. Copyright 2021 American Chemical Society.

## 2.5 What Is the Difference Between Active and Passive Transport?

As mentioned earlier, in order for a xenobiotic to enter the body it needs to pass across cell membranes. Once the chemical gets to the cell membrane there are two mechanisms by which it can enter into the cell, passively or actively.

Passive transport represents the simple diffusion of the chemical through the cell membrane without assistance or the addition of energy. This is the way that most xenobiotics cross the membrane, and it is determined by the ability of the chemical to move either through the small pores or the lipid bilayer, and the concentration gradient on both sides of the membrane. Passive transport of a chemical is determined by the size of the molecule, lipophilicity, and degree of ionization. Small water-soluble molecules ($M_W$ up to 600 Da) pass through the aqueous pores in the membrane, while larger molecules ($M_W$ up to 60,000 Da) can only pass through the pores in the specialized cells of capillaries and kidneys.[4] The rate of the diffusion of lipophilic molecules correlates well with the log P values, meaning that chemicals with high lipid solubility (high log P), easily diffuse through the phospholipid membrane.

Active transport involves the participation of special transport proteins embedded through the cell membrane (as shown in Fig. 2.6). These proteins can transport the chemical either following the concentration gradient (from high to low concentration), i.e., by assisted diffusion, or going against the concentration gradient (from low to high concentration) which requires cellular energy. Assisted diffusion helps the transfer of the larger molecules that have difficulty diffusing through the membrane via aqueous pores or the phospholipid layer. Active transport is an important mechanism in the transportation of xenobiotics to the different organs and organ systems, and the maintenance of the nutrient and electrolyte balance through the body. Figure 2.8 schematically shows the passive and active transport of chemicals through the cell membrane while Table 2.2 presents examples of chemicals that are passively and actively transported.

**Table 2.2** Examples of different types of transport through the cell membrane

| | |
|---|---|
| Passive | Small non-charged molecules such as $O_2$, $CO_2$, and $H_2O$ can pass through cell membrane freely |
| Assisted | Transport of sugar and amino acids into the CNS |
| Active | Sodium–potassium pump uses ATP to move three $Na^+$ ions and two $K^+$ to the site where there are already highly concentrated |

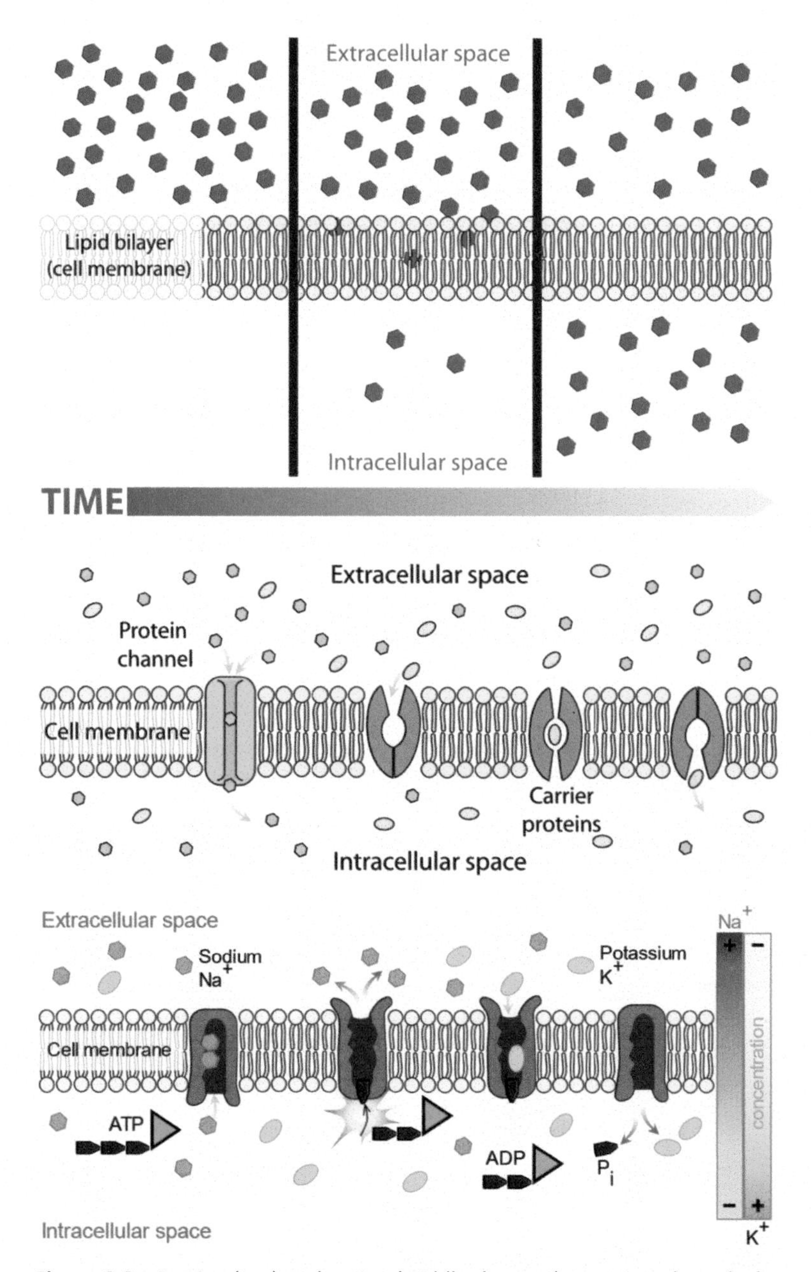


**Figure 2.8** Passive (top) and active (middle, bottom) transport through the cell membrane.[17]

## 2.6 Why Is Metabolism Crucial to Toxicity?

Metabolic transformations, i.e., metabolism, is the term used to describe the process by which a molecule gets transformed to another molecule by chemical reactions happening in the body. Thus, this process is otherwise known as biotransformation.[18] Normally, biotransformation is responsible for transforming the nutrients absorbed by ingesting food and liquid, or by breathing the air, into molecules required for the functioning of the body. However, it is also a very important process in the protection of the body from various toxic chemicals. This defense mechanism is responsible for converting the body wastes and harmful xenobiotics to less harmful substances, or it can make them easier to excrete from the body. For example, most of the xenobiotic biotransformations convert lipophilic molecules (that are easily absorbed in the GI tract and other sites) into hydrophilic molecules (that are easily excreted with urine or bile), while the methylation reaction actually makes some xenobiotics less water-soluble. Most of the harmful xenobiotics are biotransformed safely in the body, i.e., detoxified, however, there are cases where the transformation of harmful chemicals results in more toxic substances, i.e., they become bioactivated through the reactions shown in Fig. 2.9 and described in more detail in the following section.

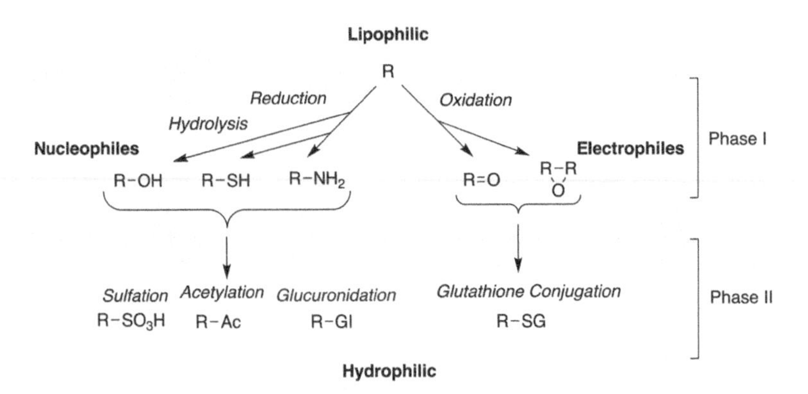


**Figure 2.9** Examples of biotransformation reactions.[19]

The major chemical reactions involved with the biotransformation are oxidation, reduction, hydrolysis, and conjugation (Fig. 2.9).

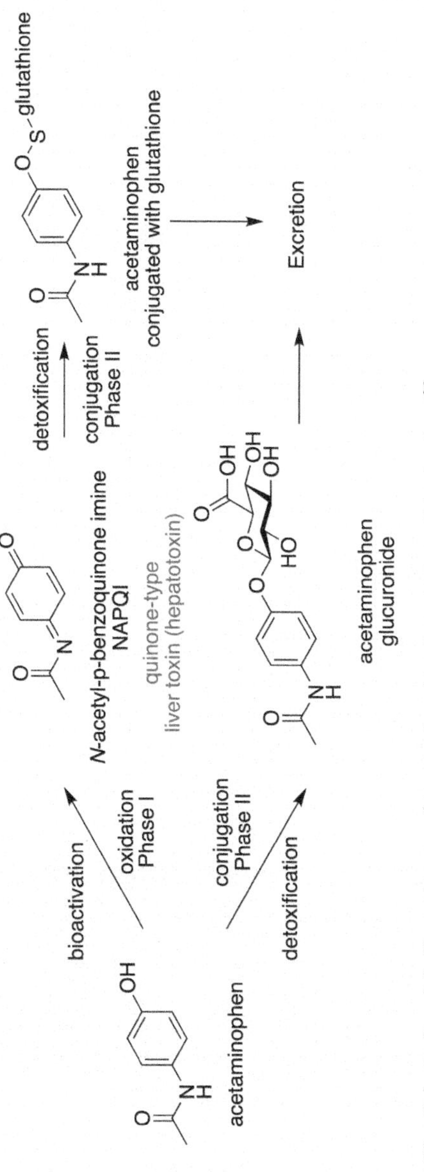


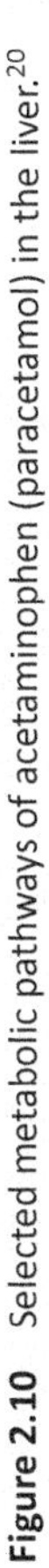

**Figure 2.10** Selected metabolic pathways of acetaminophen (paracetamol) in the liver.[20]

These reactions are catalyzed by various enzymes and are often categorized by the sequence that biotransformation occurs into phase I and phase II reactions. Phase I reactions are generally simple chemical transformations of molecules where various functional groups are introduced, while phase II reactions represent more complicated conjugations performed by enzymes. Chemicals that are biotransformed by the phase I reactions convert into either sufficiently ionized (hydrophilic) molecules that will be excreted, or into a metabolite that will undergo phase II conjugations. Sometimes, molecules will skip the phase I reactions if they already possess the functional groups that will make them suitable for phase II conjugations. Figure 2.10 shows selected metabolic pathways of acetaminophen molecule biotransformation in the liver.[20]

While biotransformation is the process that either detoxifies or bioactivates the harmful xenobiotics, its effectiveness depends on several factors that are directly connected with the exposed individual. These factors are the species, age and gender, genetic variability, nutrition, prior or simultaneous exposure to other chemicals, and finally the dose level.[4]

Most of the biotransformations take place in the liver, the organ that is the largest within the body and possesses a high concentration of biotransforming enzymes, some take place in the kidneys and lungs, and partially in the intestines, skin, and reproductive organs. Thus, these organs are often the sites where the adverse effects are mostly expressed.

## 2.7 What Is the Bioactivation Step in Metabolism?

While metabolism processes in the body exist to protect the individual, sometimes biotransformation reactions can result in the increased toxicity of the transformed chemical. This process is called bioactivation, and it is a basis for the toxicity of a large number of xenobiotics that are otherwise not so reactive.[21] More reactive and more toxic metabolites are generally created by phase I reactions; however, phase II reactions can result in the bioactivation of chemicals either by themselves or in combination with phase I reactions. The majority of these reactive metabolites can still be detoxified, so toxic effects will exhibit themselves only when the capacity for detoxication is diminished, or if their formation is enhanced. Most of the bioactivation reactions involve transformations by cytochrome P450 (CYP450) and peroxidase enzymes to close-shell electrophilic molecules (primarily generated by CYP450 system), and free radicals (primarily produced by peroxidases).[21]

These transformations can be classified based on the types of reactive intermediates and their potential reactivity into four categories:

(a) Biotransformation to stable but toxic metabolites: only a limited number of chemicals are at the same time stable and toxic, so this mechanism is not as common in the body. For example, dihalomethanes are oxidized by the CYP450 to carbon monoxide which binds to hemoglobin interfering with the oxygen transport (Fig. 2.11). Another illustrative example is the peripheral nerve injury caused when n-hexane gets oxidized to cyclic pyrrole in several steps, resulting in the changes of the 3-D structure of proteins in the neuronal protective coating.[22]

(b) Biotransformation to electrophiles: A large number of chemicals exhibit cytotoxicity or carcinogenicity by forming electrophiles that will be involved in the alkylation or acylation of lipids, proteins, or DNA molecules. As the most common mechanism of bioactivation, reactive intermediates include a broad range of chemical moieties such as acyl halides, carbocations, epoxides, nitrenium ions, and quinones. Most of these species result from oxidation mediated by the CYP450, but other enzymes can catalyze their formation as well. For example, vinyl chloride is transformed by CYP450 to electrophilic oxiranes that are linked to carcinogenicity.[24] Glutathione conjugates 1,2-dibromoethane to S-(2-bromoethyl)glutathione, which transforms into the highly reactive episulfonium ion (Fig. 2.12).

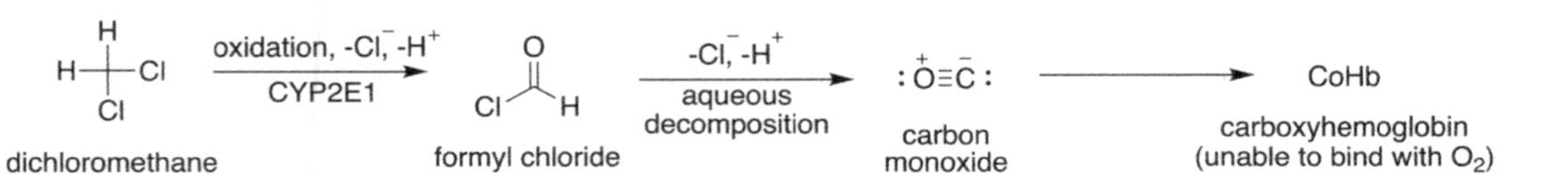


**Figure 2.11** Biotransformation of dichloromethane to carbon monoxide that binds to hemoglobin.[23]

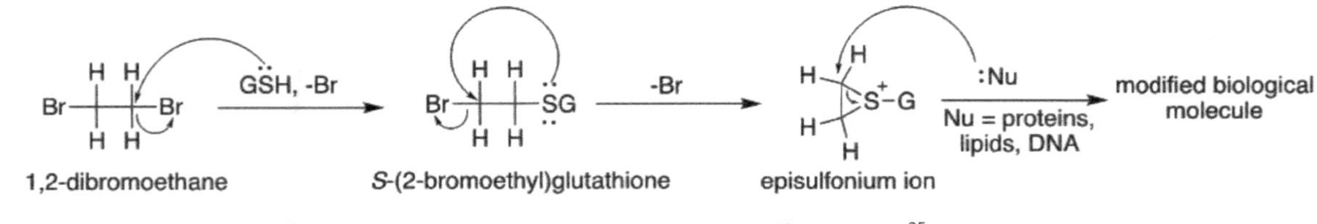


**Figure 2.12** Biotransformation of 1,2-dibromoethane to highly reactive episulfonium ion.[25]

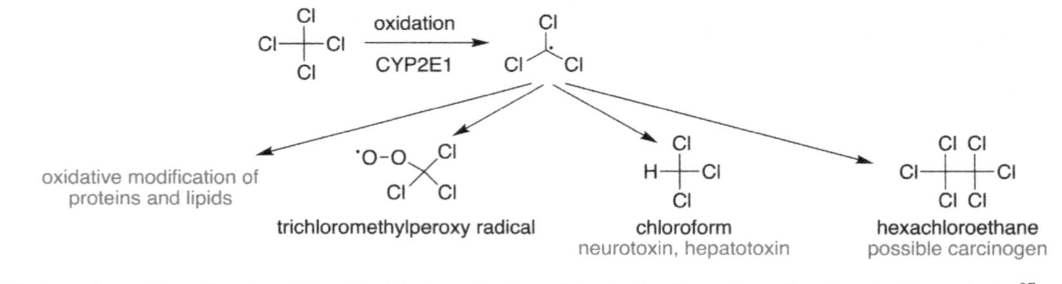

**Figure 2.13** Biotransformation of carbon tetrachloride to radical resulting in the formation of various toxic products.[27]

(c) Biotransformation to free radicals: free radicals are highly reactive species that can be formed by one-electron oxidation to give a cation radical. These radicals can undergo oxidation–reduction reactions, abstract hydrogen atoms, result in dimerization or disproportionation reactions, and participate in chain reactions. One of the most representative examples is the bioactivation of carbon tetrachloride to a radical form (Fig. 2.13). The trichloromethyl radical can abstract hydrogen from macromolecules releasing chloroform or dimerize giving hexachloroethane.[26] Toxic effects caused by radicals are exhibited as the oxidative modification to proteins and lipid peroxidation.

(d) Formation of reactive oxygen species: Oxidative stress, i.e., cell damage, is the mechanism that involves reduced oxygen species such as hydrogen peroxide, the hydroxy radical, and the superoxide radical anion. Certain xenobiotics are able to undergo redox cycling or enzymatic oxidoreductions resulting in these reactive species. For example, during the quinone redox cycling a superoxide radical anion is formed, which is then transformed to hydrogen peroxide, and finally a highly reactive hydroxyl radical (Fig. 2.14). The hydroxyl radical can then initiate cell damage.

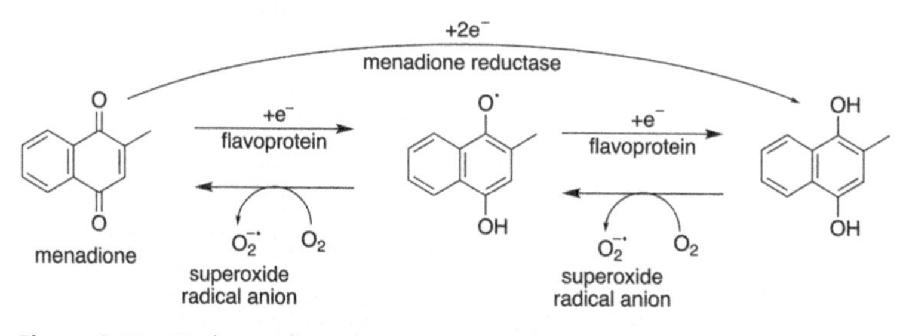


**Figure 2.14** Redox cycling of quinones resulting in the formation of highly reactive hydroxyl radical.

## 2.8 What Is the Detoxification Step in Metabolism?

The natural response of the body to any toxic xenobiotic is to metabolize it in such a way that the toxic effect will be neutralized, and the xenobiotic transformed into an easy to excrete form. Several mechanisms exist that are responsible for the detoxification of various molecular species.[28]

Free radicals such as the hydroxyl radical ($HO^{\bullet}$), carbonate anion ($CO_3^{\bullet-}$), and nitrogen dioxide ($^{\bullet}NO_2$) are usually generated from the superoxide radical ($O_2^{\bullet-}$). Thus, superoxide dismutase enzymes located in mitochondria and cytosol are responsible for converting $O_2^{\bullet-}$ to hydrogen peroxide which is further reduced to water by peroxisomal catalase or glutathione peroxidase enzymes. This is especially important for the elimination of $HO^{\bullet}$ because it has a very short half-life ($10^{-9}$ s) and no enzyme actively removes this radical from the body. Glutathione peroxidase also reduces the dangerous peroxynitrite radical ($ONOO^-$) that reacts with oxyhemoglobin, to nitrite ($ONO^-$) as shown schematically in Fig. 2.15.

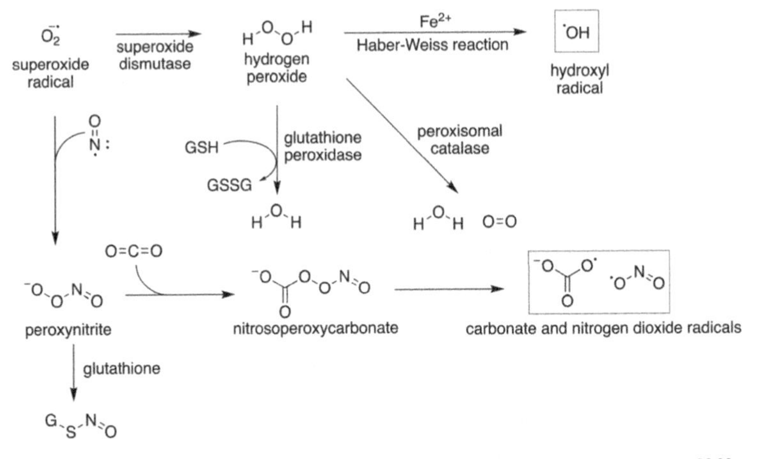

**Figure 2.15** Reactions of free-radical formation and their detoxification.[28,29]

Electrophiles are detoxified by conjugation with glutathione, a good nucleophile, which can happen spontaneously or by activation with glutathione S-transferase enzymes (Fig. 2.16).[28] On the other hand, nucleophiles are usually detoxified by conjugation of a functional group to the nucleophilic atom by common reactions such as acetylation, glucuronidation, methylation, and sulfonation. These conjugations are responsible for the prevention of aminophenol, catechol, hydroquinone, and phenol transformations to toxic electrophilic quinines and quinoneimines.

Other toxicants, such as chemicals without distinctive functional groups (e.g., benzene, see Fig. 2.17), are detoxified in the previously mentioned phase I and phase II reactions.[28,30] First, functional groups such as the carboxyl or hydroxyl group are introduced, usually by CYP450 enzymes. Then, transferase adds some endogenous acid, such as amino acid, glucuronic acid, or sulfuric acid, to the functional group resulting in a highly hydrophilic organic acid that is inactive, and easily excreted.

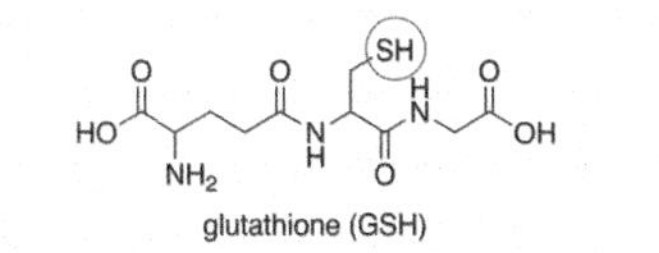


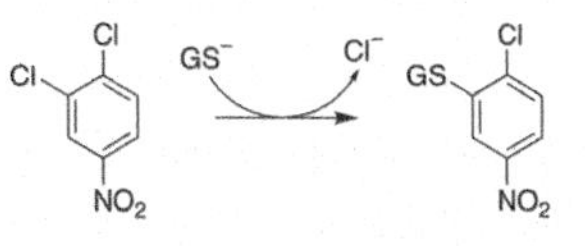


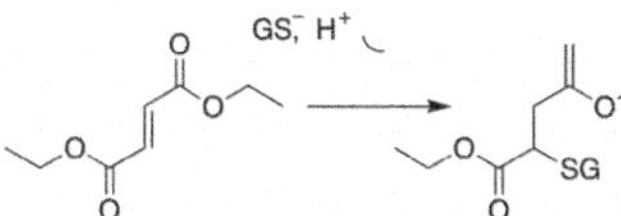


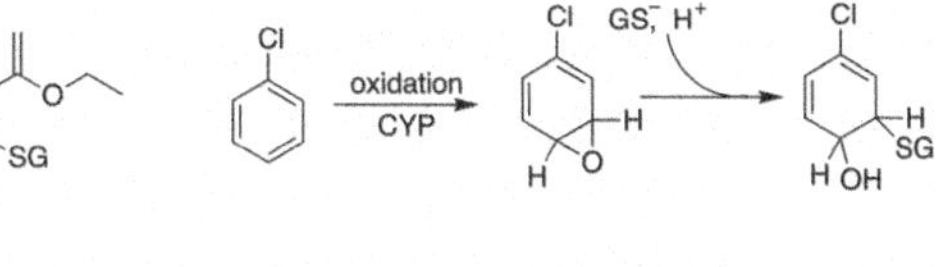


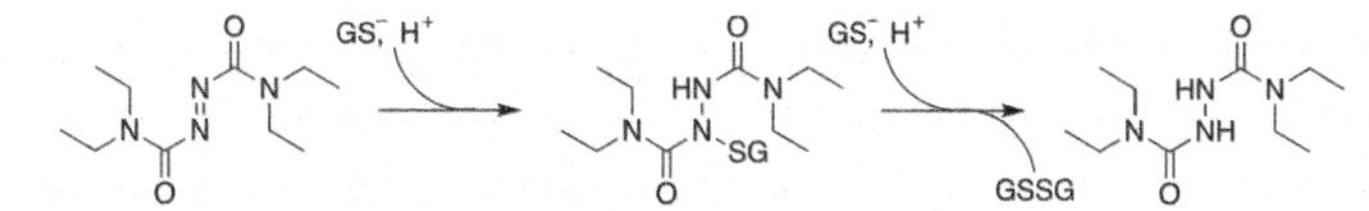


**Figure 2.16** Selected electrophiles detoxification reactions.

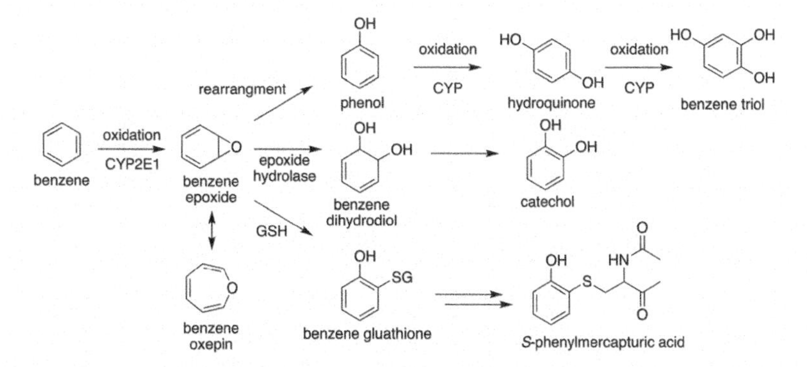


**Figure 2.17** Benzene detoxification reaction pathways.[30]

Some toxins, originating from natural venoms, are proteins that are inactivated by intra- and extracellular protease enzymes. For example, the enzyme thioredoxin is responsible for inactivating α- and β-bungarotoxin, phospholipase, and erabutoxin b, by reducing intramolecular disulfide bonds found in these proteins.[4]

Sometimes, detoxification can fail because the concentration of the toxic chemical overwhelms the detoxification process, or the enzyme responsible for the process gets inactivated by a very reactive molecule that covalently binds to it. Additionally, detoxification processes can result in harmful by-products, or the successfully detoxified chemical may be bioactivated at some other site in the body.

## 2.9 What Is Distribution and Why Is It Crucial to Toxicity?

Once the xenobiotic gets absorbed into the body, via different routes of exposure described earlier, it first gets into the interstitial fluids, i.e., those surrounding cells, and then to the intercellular fluids or blood plasma. The distribution of the xenobiotic continues by migration to the cardiovascular and lymphatic system, where the blood or lymph carries them to their site(s) of action. Once the chemical gets to the bloodstream it can get excreted from the body via different organs, deposited in various cells and tissues, or biotransformed into different chemicals. The overall distribution of toxic chemicals through the body is shown schematically in Fig. 2.18.

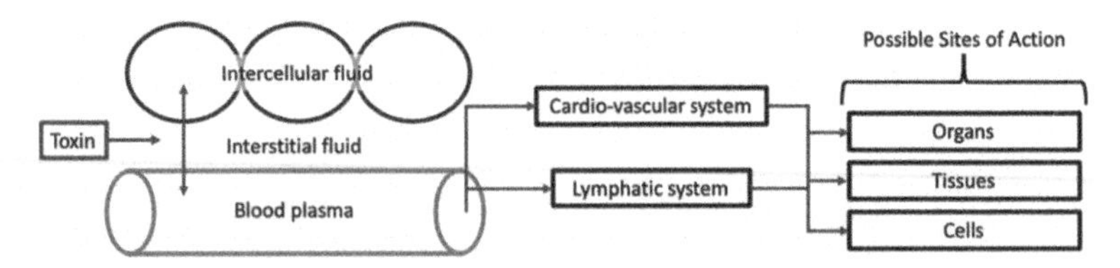


**Figure 2.18** Overall distribution of toxins within the body.

The concentration of the toxic chemicals or their metabolites at any site is very much dependent on the route of exposure. This will determine the path of a chemical as it moves through the body and the time it spends at a certain site, which will influence the level of biotransformation, storage, and excretion of said chemical. In toxicology, the distribution of a chemical in body fluids is usually estimated by calculating the apparent volume of distribution ($V_d$) that can be derived from the following equation:

$$V_d = \frac{\text{dose (mg)}}{\text{plasma conc.}\left(\frac{\text{mg}}{\text{L}}\right)}$$

A high value of $V_d$ may indicate the toxic chemical has been distributed to a certain organ or tissue.

Whether the chemical will be easier or harder to distribute depends on several factors. These include factors influencing the passage through the cell membranes, i.e., molecular weight, lipid solubility, polarity of the molecule, and concentration gradient. Additionally, the xenobiotic can bind to the proteins present in blood plasma which affects their half-life within the body.

Organs and tissues in the body are exposed to different amounts of xenobiotics because there is a difference in the volume of blood flowing through specific tissues, and special barriers that slow the entrance of toxic chemicals exist to protect certain organs. The tissues can have a high affinity for some chemicals, and this affinity will influence the concentration of a chemical in that tissue. Additionally, the high flow of blood in certain organs (brain, lungs, kidneys, and heart), leaves these organs exposed to increased amounts of toxicants. The barriers that protect the brain, placenta, and testes are structural barriers consisting of specialized cells (or layers of cells) that limit the diffusion of toxicants while allowing the transport of nutrients essential for the functioning of these organs.

As mentioned earlier, distribution of the xenobiotics through the organism can result in their accumulation and storage in tissues like bone, where elements such as strontium or lead can substitute for calcium (due to their similar properties because of the same oxidation states); liver and kidney tissues, which are especially prone to the chemicals binding to their proteins; and adipose tissue, where very lipophilic molecules can accumulate.

## 2.10 What Properties Promote Distribution?

As we mentioned in the previous section, the distribution of a chemical depends on several factors that influence the solubility, passage through the cell membranes, and tendency to bind to certain proteins existing in the inter- and intracellular space.

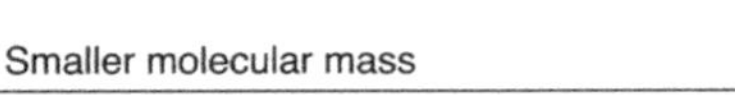

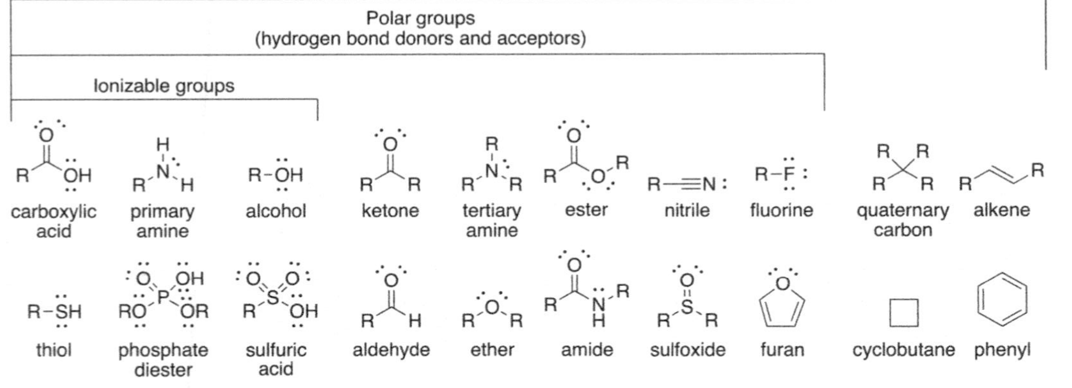


**Figure 2.19** Factors and functional groups that promote distribution.

The aqueous solubility and lipophilicity of some chemical compound have a big influence on their distribution through body fluids and their passage through various biological barriers. Lipophilicity generally influences the passage of molecules, i.e., permeability, through biological barriers such as the blood-brain barrier, which consists of a network of thin-walled capillaries. Having higher lipophilicity will allow a molecule to pass through the barriers without too much difficulty, while higher aqueous solubility means that a molecule will tend to remain in the extracellular space or blood more often.[31]

Aqueous solubility will increase if there are ionizable groups present, such as an amino or carboxylic group; reducing the lipophilic character, i.e., log P value, will result in increased solubility and overall systemic exposure. Aqueous solubility can also be increased by the addition of polar groups, or hydrogen bond donors and acceptors, such as -OH and -$NH_2$ groups. The size of a molecule influences the solubility as well, as smaller molecules with lower $M_w$ are generally more soluble and metabolically stable than larger molecules. The molecule's capacity to permeate is increased with the addition of more lipophilic groups, e.g., longer nonpolar hydrocarbon chains; by esterification of carboxylic acid; replacing the ionizable group with non-ionizable groups; reducing the size of the molecule, or by reducing the hydrogen bonding (Fig. 2.19).

The physicochemical properties influencing the solubility and permeability are in actuality intercorrelated, and often structural changes that increase aqueous solubility will reduce permeability. This means that the molecules with properly balanced properties will have the highest distribution rate throughout the whole body.

## 2.11 What Is Excretion and Why Is It Crucial to Toxicity?

The elimination of harmful xenobiotic chemicals from the body is a very important process that naturally occurs via specialized excretory organs. In order to get out of the body, chemicals need to pass through various cell membranes and tissues, which will be affected by the same physicochemical properties that determine the rate of absorption and distribution described earlier.[31] This generally means that hydrophilic molecules are more likely to be excreted compared to lipophilic molecules. The main routes the body employs to remove toxic chemicals, or their metabolites, are via exhaled air, feces, and urine (Fig. 2.20). A small amount of chemicals can be excreted by breast milk, sweat, or other body secretions, but these are only minor routes.

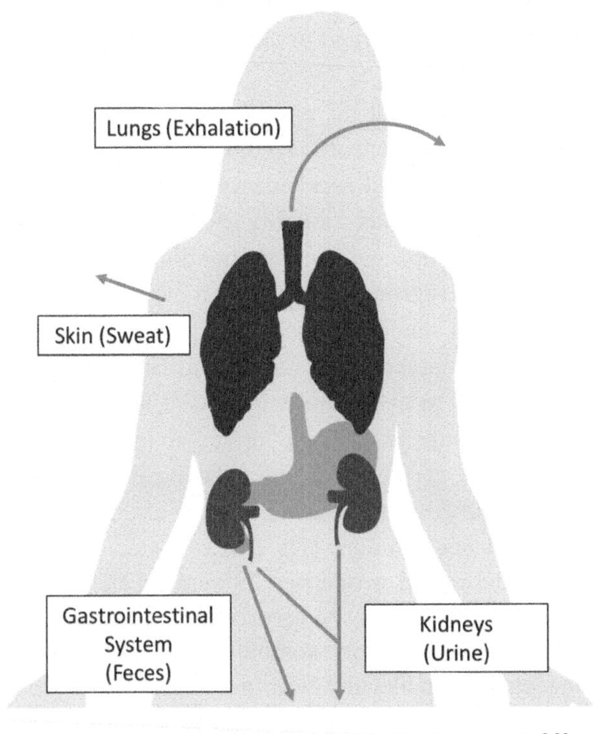

**Figure 2.20** Typical ways chemical gets excreted from the body.[2,32]

The kidneys are the organ through which most of the toxic substances are eliminated from the body. Specialized areas in the kidneys, i.e., nephrons, consisting of a network of interconnected capillaries, are responsible for filtration, secretion, and reabsorption of water-soluble waste chemicals and toxicants. These molecules get collected and transferred from the blood to the urine and subsequently removed from the body via the bladder.

The gastrointestinal tract is responsible for the excretion of waste products of the body functions through feces. Additionally, harmful molecules can also be eliminated from the body in such a way either by direct excretion into the intestine or via the bile that is excreted by the liver to the intestinal space. Since most biotransformation reactions occur in the liver, xenobiotics and their metabolites are mainly excreted in the form of bile. Intestinal excretion occurs very slowly by passive diffusion of xenobiotics through the capillary's walls directly into the intestinal space.

Xenobiotics existing in the gaseous form in the blood are removed by the lungs via passive diffusion. This excretion mechanism occurs when the concentration of the gaseous molecule dissolved in the capillary blood exceeds the concentration of this molecule in the alveolus space. If the solubility of a gaseous or liquid substance in the blood is low, it will be readily excreted by the lungs with expired air. Additionally, because the capillaries are so close to the thin alveolar membranes, even lipophilic substances can be efficiently removed by the lungs.[4]

## References

1. Lehman-McKeeman, L. D. Absorption, distribution, and excretion of toxicants. In: Klaassen, C. D.; Watkins, III, J. B. (Eds.), *Casarett & Doull's Essentials of Toxicology*, 3rd ed., McGraw-Hill Education: New York, 2015, pp. 61–77.
2. (a) infographic dalaman sistem organ (changes made, CC0 1.0). https://svgsilh.com/ms/3f51b5/image/37824.html (accessed April 30); (b) internal system organ anatomy (changes made, CC0 1.0). https://svgsilh.com/ffeb3b/image/310730.html (accessed April 30); (c) anatomy medicine female skin (changes made, CC0 1.0). https://svgsilh.com/3f51b5/image/876199.html (accessed April 30).
3. Cheremisinoff, N. P.; Rosenfeld, P. E. (Eds.) Chapter 6: DDT and related compounds. *Handbook of Pollution Prevention and Cleaner Production: Best Practices in the Agrochemical Industry*, William Andrew Publishing: Oxford, 2011, pp. 247–259.
4. Klaassen, C. D.; Watkins, III, J. B., *Casarett & Doull's Essentials of Toxicology*, 3rd ed., McGraw-Hill Education: New York, 2015.
5. (a) respiratory illness smoke lip (changes made, CC0 1.0). https://svgsilh.com/2196f3/image/156101.html (accessed April 30); (b)

design medical science anatomy (changes made, CC0 1.0). https://svgsilh.com/ffeb3b/image/304086.html (accessed April 30).

6. National Institute for Occupational Safety and Health. Skin exposures and effects. https://www.cdc.gov/niosh/topics/skin/default.html (accessed December 1).
7. Clark, M. A.; Douglas, M.; Choi, J. *Biology 2e*. OpenStax: Houston, Texas, 2018, Access for free at https://openstax.org/books/biology-2e/pages/1-introduction.
8. Lipinski, C. A.; Lombardo, F.; Dominy, B. W.; Feeney, P. J. Experimental and computational approaches to estimate solubility and permeability in drug discovery and development settings. *Adv Drug Deliver Rev*, 1997, **23**(1–3), 3–25.
9. Lipinski, C. A. Lead- and drug-like compounds: the rule-of-five revolution. *Drug Discov Today Technol*, 2004, **1**(4), 337–341.
10. Kerns, E. H.; Di, L. Pharmaceutical profiling in drug discovery. *Drug Discov Today*, 2003, **8**(7), 316–323.
11. Ghose, A. K.; Viswanadhan, V. N.; Wendoloski, J. J. A knowledge-based approach in designing combinatorial or medicinal chemistry libraries for drug discovery. 1. A qualitative and quantitative characterization of known drug databases. *J Com Chem*, 1999, **1**(1), 55–68.
12. Veber, D. F.; Johnson, S. R.; Cheng, H.-Y.; Smith, B. R.; et al. Molecular properties that influence the oral bioavailability of drug candidates. *J Med Chem*, 2002, **45**(12), 2615–2623.
13. Jhoti, H.; Williams, G.; Rees, D. C.; Murray, C. W. The 'rule of three' for fragment-based drug discovery: where are we now? *Nat Rev Drug Discov*, 2013, **12**(8), 644–644.
14. National Research Council. *A Framework to Guide Selection of Chemical Alternatives*. The National Academies Press: Washington, DC, 2014.
15. Moldoveanu, S.; David, V. Phase transfer in sample preparation. In: Moldoveanu, S.; David, V. (Eds.), *Modern Sample Preparation for Chromatography*, Elsevier: Amsterdam, 2015, pp. 105–130.
16. Stenberg, P.; Norinder, U.; Luthman, K.; Artursson, P. Experimental and computational screening models for the prediction of intestinal drug absorption. *J Med Chem*, 2001, **44**(12), 1927–1937.
17. (a) Villarreal, M. R. Scheme simple diffusion in cell membrane. https://commons.wikimedia.org/wiki/File:Scheme_simple_diffusion_in_cell_membrane-en.svg (accessed January 15); (b) Villarreal, M. R. Scheme facilitated diffusion in cell membrane. https://commons.wikimedia.org/wiki/File:Scheme_facilitated_diffusion_in_cell_membrane-en.svg

(accessed 15 January); (c) Villarreal, M. R. Scheme sodium-potassium pump. https://commons.wikimedia.org/wiki/File:Scheme_sodium-potassium_pump-en.svg (accessed January 15).

18. Parkinson, A.; Ogilvie, B. W.; Buckley, D. B.; Kazmi, F.; et al. Biotransformation of xenobiotics. In: Klaassen, C. D.; Watkins, III, J. B. (Eds.), *Casarett & Doull's Essentials of Toxicology*, 3rd ed., McGraw-Hill Education: New York, 2015, pp. 79–107.
19. Vickers, T. Outline of phase I and II of xenobiotic metabolism. https://commons.wikimedia.org/wiki/File:Xenobiotic_metabolism.png (accessed October 1).
20. Mazaleuskaya, L. L.; Sangkuhl, K.; Thorn, C. F.; FitzGerald, G. A.; et al. PharmGKB summary: pathways of acetaminophen metabolism at the therapeutic versus toxic doses. *Pharmacogenet Genomics*, 2015, **25**(8), 416–426.
21. Dekant, W. The role of biotransformation and bioactivation in toxicity. In: Luch, A. (Ed.), *Molecular, Clinical and Environmental Toxicology, Volume 1: Molecular Toxicology*, Birkhäuser Verlag: Basel, 2009, pp. 57–86.
22. Spencer, P. S. Neuroprotein targets of γ-diketone metabolites of aliphatic and aromatic solvents that induce central–peripheral axonopathy. *Toxicol Pathol*, 2020, **48**(3), 411–421.
23. (a) United States Environmental Protection Agency. Toxicological review of dichloromethane (methylene chloride). U.S. EPA: Washington, DC, 2011, p. 567; (b) Phillips, D. L.; Zhao, C.; Wang, D. A Theoretical study of the mechanism of the water-catalyzed HCl elimination reactions of CHXCl(OH) (X = H, Cl) and HClCO in the gas phase and in aqueous solution. *J Phys Chem A*, 2005, **109**(42), 9653–9673.
24. Bond, J. A.; Melnick, R. L. Electrophilic agents. In: Baan, R. A.; Stewart, B. W.; Straif, K. (Eds.) *Tumour Site Concordance and Mechanisms of Carcinogenesis*, IARC Scientific Publications (165th ed.), International Agency for Research on Cancer: Lyon (FR), 2019.
25. United States Environmental Protection Agency. Toxicological review of 1,2-dibromoethane, U.S. EPA: Washington, DC, 2004, p. 240.
26. United States Environmental Protection Agency. Toxicological review of carbon tetrachloride, EPA/635/R-08/005F, U.S. EPA: Washington, DC, 2010.
27. Agency for Toxic Substances and Disease Registry. Carbon tetrachloride toxicity. Environmental Health and Medicine Education, U.S. Department of Health and Human Services, United States Centers for Disease Control and Prevention, 2017, p. 79.

28. Klaassen, C. D.; Watkins, III, J. B. *Casarett & Doull's Essentials of Toxicology*, 3rd ed., McGraw-Hill Education: New York, 2015, p. 524.

29. Jacob-Ferreira, A. L.; Schulz, R. Activation of intracellular matrix metalloproteinase-2 by reactive oxygen–nitrogen species: Consequences and therapeutic strategies in the heart. *Arch Biochem Biophys*, 2013, **540**(1), 82–93.

30. United States Environmental Protection Agency. *Toxicological Review of Benzene (Noncancer Effects)*, United States Environmental Protection Agency: Washington, DC, 2002, p. 180.

31. Kerns, E. H.; Di, L. *Drug-like Properties: Concepts, Structure Design and Methods*, Elsevier: Burlington, MA, 2008, p. 526.

32. organ anatomy biology human, (changes made, CC0 1.0). https://svgsilh.com/3f51b5/image/158998.html (accessed April 30).

# Chapter 3

# Degradability

## 3.1 What Is the Difference Between Degradability and Biodegradability?

The degradability of a chemical substance is the capability of the said chemical to decompose under certain conditions. In nature, most substances and molecules are not permanently stable but, given enough time and right conditions can decompose into smaller and chemically distinct fragments. There are two main ways this can happen, either under the influence of different environmental conditions, i.e., abiotically; or by the work of microorganisms, i.e., biotically (see Fig. 3.1).[1] Abiotic degradation will cause chemical decomposition of molecules when they are exposed to sufficient heat (thermolysis), light (photolysis), or water (hydrolysis).[2] The conditions and specifics for each of these processes will be discussed in the following sections. While the chemicals can decompose without the influence of any microorganism, once they get into the environment a large number of species of bacteria, fungi, and yeasts can absorb and transform these chemicals.

The biodegradability of chemical compounds depends not only on their molecular structure and physicochemical properties but also on the broad range of conditions existing in the environment they are exposed to. These include the presence of different microorganisms,

*First Do No Harm: A Chemist's Guide to Molecular Design for Reduced Hazard*
Predrag V. Petrovic and Paul T. Anastas

ISBN 978-981-4968-59-1 (Hardcover), 978-1-003-35964-7 (eBook)
www.jennystanford.com

the existence of adequate nutrients for the microbial population, concentration of the chemical, temperature, pH, and covalent binding of the chemical to other substrates in the environment (e.g., humus, clay). Obviously, some of these factors are not as important from the molecular design perspective since they cannot be modified actively. Generally, a higher temperature (within the normal environmental condition) will promote biodegradation, as well as a nutrient-rich environment, but these cannot be influenced by better design. On the other hand, water-soluble molecules are generally better biodegraded, though many microorganisms secrete their own biosurfactants, increasing the solubility rate of the lipophilic molecules.[3] Once chemicals get into the cells, their biodegradation rate will depend on their molecular structure and physicochemical properties, as described earlier.

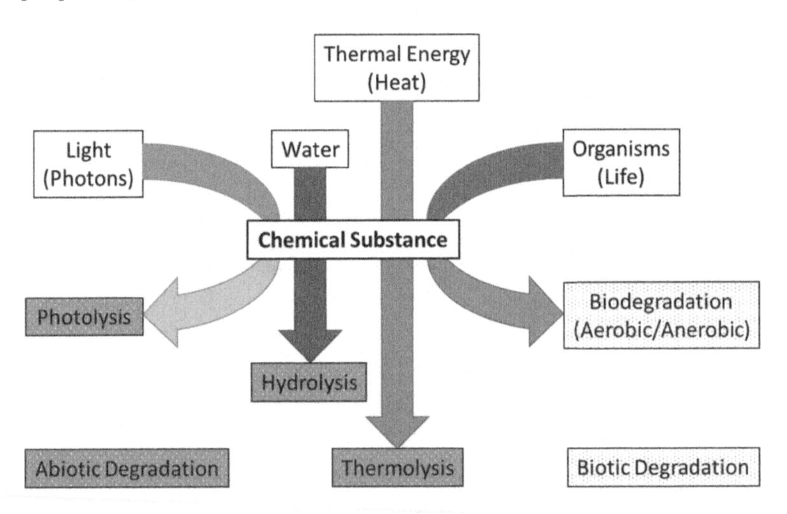


**Figure 3.1** Schematic representation of different kinds of degradation.

## 3.2 What Is Photolytic Degradation?

Photodegradation, i.e., photolysis—is a chemical reaction in which a molecule is broken down by one or more photons possessing sufficient energy to break chemical bonds in the molecule. This reaction is usually initiated when infrared, visible, and ultraviolet light get absorbed by the atom or molecule. This increases the energy of the atom or molecule, exciting it to a transition state that changes its physicochemical properties. A molecule in this state can change its structure, combine with other molecules, transfer atoms, electrons, protons, or energy to other molecules causing a chain reaction (Fig. 3.2).

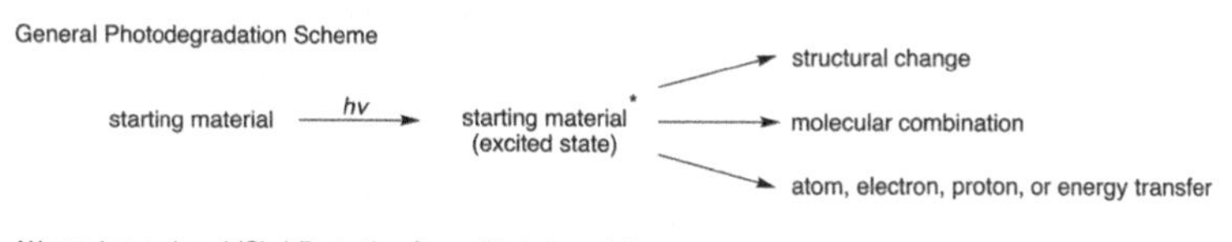


N-benzyloxycarbonyl (Cbz) Protection Group Photodegradation

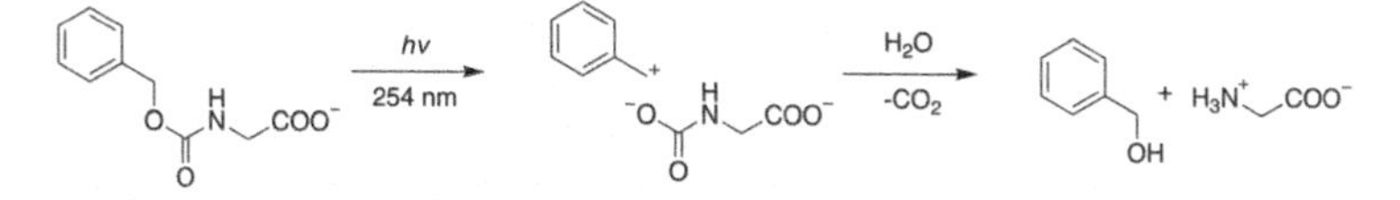


Norrish Type II Photodegradation of Alky Phenyl Ketone

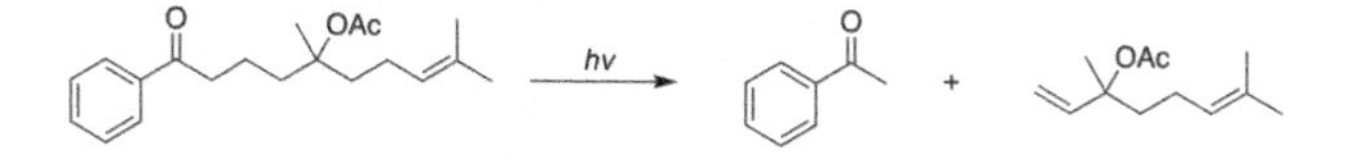


Photoinitiated Radicals of Polystyrene

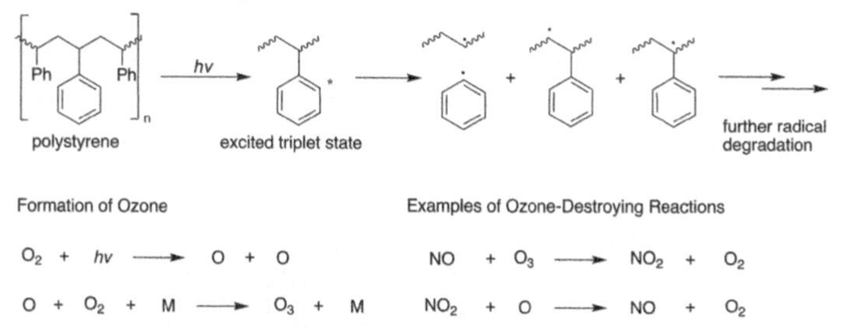


**Figure 3.2** Selected photolytic degradation reactions.[4]

The photolytic degradation of a molecule can be promoted by the presence of photolabile functional groups (Fig. 3.3).[4b] These functional groups possess weak bonds that can be dissociated, i.e., their bond dissociation energies are smaller than the absorbed UV radiation. Additionally, aromatic rings, heteroatoms, unsaturated bonds, or hydroperoxide allow for the absorption of UV radiation that can initiate photodegradation reactions.[4d] For example, epoxy and hydroxyl groups were found to be directly responsible for the increased photodegradation rates of graphene oxide nanomaterials.[5]

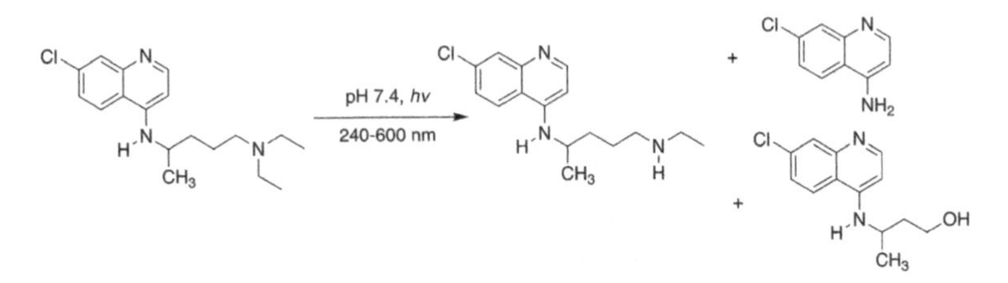


**Figure 3.3** Photolytic degradation of immunosuppressive and anti-parasite drug chloroquine.

The same nature of photolabile groups makes them very useful as protective groups in organic synthesis. For example, photolabile ester groups are used to mask the carboxylic moiety and release carboxylic acids, aldehydes, and ketones.[6]

One of the extremely important photolytic reactions is the genesis of the ozone molecule in the stratosphere. Ozone is created when UV radiation strikes an oxygen molecule splitting it into individual atoms that then recombine into ozone ($O_3$) (Fig. 3.2). When this reaction happens high in the stratosphere it builds the ozone layer, which is responsible for absorbing short wavelengths of harmful UV-B and UV-C radiation. However, UV radiation can also dissociate nitrogen molecules that transform to nitric oxide (NO) and nitrogen dioxide ($NO_2$) which can then act as a catalyst in the destruction of the ozone (Fig. 3.2).

## 3.3 What Is Hydrolytic Degradation?

Hydrolysis is an important abiotic transformation process for certain classes of synthetic chemicals that are released in both freshwater and marine aquatic environments.[7] While hydrolysis of these compounds take various pathways, other chemical structures are inert. In hydrolysis reactions, a water molecule splits into hydrogen and a hydroxy group that can each (or both) react with a chemical, usually with the help of an acid or base catalyst. In the environment, this reaction is affected by the ambient temperature, pH, ionic strength, and presence of other substances.

Hydrolysis can be represented by a reversible equation:

$$AB + HOH \rightleftharpoons AH + BOH$$

where the products of the reaction can be either neutral or ionic molecules.

In a common type of hydrolysis, water spontaneously ionizes into hydroxide anions ($OH^-$) and hydronium cations ($H_3O^+$) that react with salts of weak acids or bases dissolved in the water. However, strong acids such as sulfuric acid can also hydrolyze giving a hydronium cation and bisulfate anion ($HSO_4^-$). The typical reaction of hydrolysis is saponification, where esters react with sodium hydroxide forming an alcohol and salt of carboxylic acid, e.g., triglyceride gets converted to glycerol and salts of constituent fatty acids.

The reactivity of a chemical is the most important factor when assessing the hydrolysis potential. Some substituents are readily displaced from molecules by hydrolysis reactions since they form products that are very stable in water. For example, for polymers containing various carbonyl functional groups, the hydrolysis rate will get lower following the trend: anhydride > ester > amide > ether.[8] Since the hydrolytic reactions are de facto happenings in water, the rate of hydrolysis will depend on the aqueous solubility of the molecule. The lower it is, i.e., the more hydrophobic a molecule is, the lower the rate of hydrolysis will be. Molecular weight and size also play an important role. As molecules get bigger, the rate of hydrolysis is lowered. Additionally, electron-withdrawing substituents will create partially positive centers on a molecule that will act as favorable sites for hydrolysis.

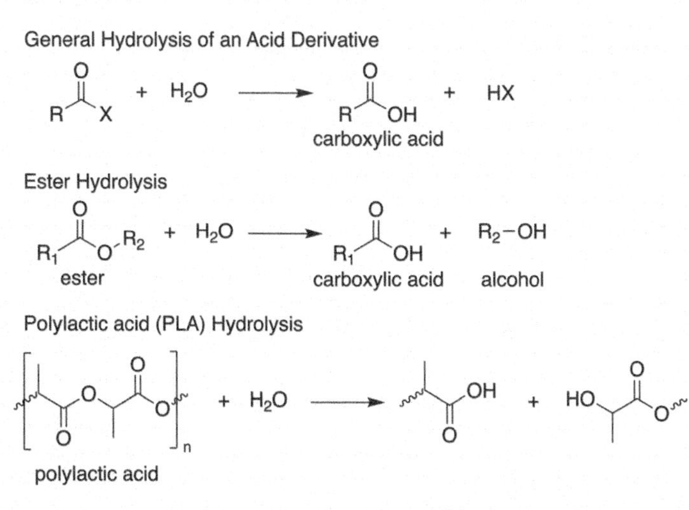


**Figure 3.4** Hydrolytic degradations of polymers.[7,9]

Hydrolysis is an important mechanism in the degradation of polymers since they can contain hydrolysable bonds, such as ester bonds in the bio-based epoxy resins (Fig. 3.4).[9]

## 3.4 What Are Thermolytic Degradation and Pyrolysis?

Thermal degradation, i.e., thermolysis—is a chemical decomposition of molecules caused by heat. This reaction is an example of an endothermic process, where the thermal energy is constantly being absorbed from the environment, resulting in the breakdown of a chemical compound into multiple different chemical substances. Thermal energy is directly related to the movement of atoms in a molecule (or molecules in a material). If sufficient energy is introduced to the molecule, the movement of its constituent atoms can be so intense that it will cause the breakage of chemical bonds, leading to the formation of distinct chemical species (Fig. 3.5). If the temperature the organic chemical is exposed to is high enough, and oxygen is absent, decomposition of the molecule is called pyrolysis.[10] The result of this kind of decomposition is the mineralization of an organic material mostly to liquids, solid char, and non-condensable gases ($H_2$, $CO$, $CO_2$, $N_2$, $CH_4$). In extreme cases pyrolysis will result in carbonization, i.e., the residue will mostly consist of carbon.

As with photodegradation, the strength of the chemical bond will influence the rate of thermal decomposition, i.e., weaker bonds will be easier to break due to the increased vibrations of constituent atoms. Once again, the size of the molecule will affect the decomposition rate, with larger molecules being more stable than smaller ones. The branching of the hydrocarbon chain can also affect the rate of thermal decomposition. For example, the thermal stability of polyolefins is strongly affected by branching with linear polyethylene being more stable, and those with branching less stable.[11] Electron-withdrawing substituents can decrease the thermostability of molecules, as in the case of substituted aromatic azo compounds.[12]

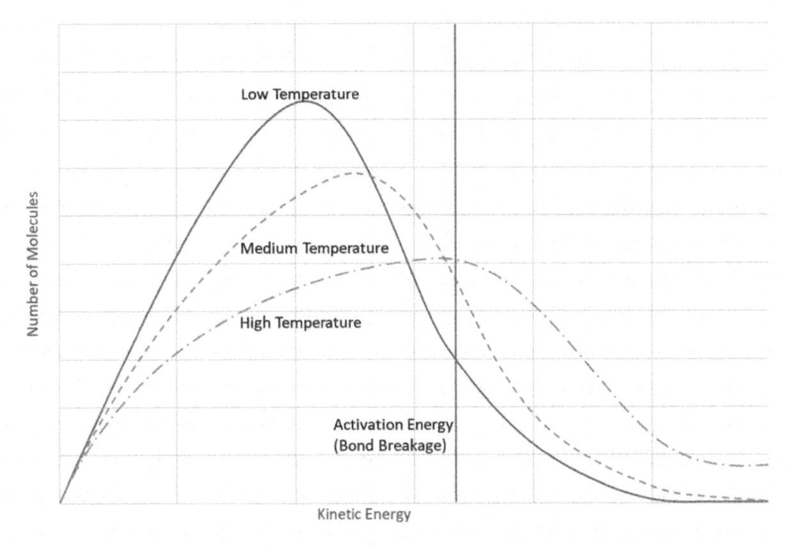


**Figure 3.5** Boltzmann distribution graph demonstrating that increasing temperature provides more molecules with sufficient energy to decompose.

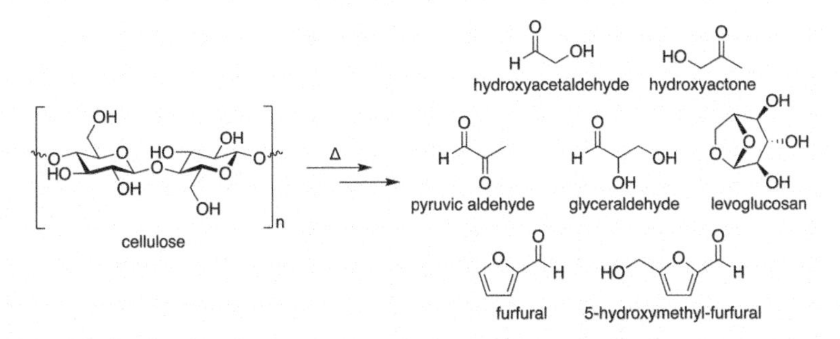


**Figure 3.6** Probable products for the thermal decomposition reactions of cellulose.

For example, when sodium bicarbonate is heated, it decomposes into carbon dioxide, water, and sodium carbonate; heating of copper(II)carbonate results in decomposition to copper(II)oxide and carbon dioxide; most metal hydroxides decompose when heated forming oxide and water. Figure 3.6 shows possible products for when cellulose is subjected to thermal decomposition.

## 3.5 How Is Biodegradability Measured?

Various methods have been used to measure the biodegradation rates over the years, however, environmentally realistic rates can be very difficult to obtain.[13] This is due to the differences in the laboratory conditions used for the estimation of biodegradation rates, and the realistic biological, chemical, and physical factors encountered in the environment. Usually, the biodegradability of organic matter is assessed using respirometry measuring the production of carbon dioxide or consumption of oxygen by the aerobic microorganisms that are involved in the degradation of the chemical. This is done by placing the chemical in a container along with the microorganisms, aerating it with oxygen or air, and measuring the amount of carbon dioxide released by the microorganisms over a set period of time. In the case where anaerobic microbes are used to estimate biodegradability, released methane is measured. Some methods also measure the loss of dissolved organic carbon for water-soluble substances, or the loss of hydrocarbon infrared absorption bands. However, since microorganisms cannot degrade all the chemicals in a reasonably short amount of time, the biodegradability of these chemicals can only be approximated. As mentioned previously, the rate of biodegradation can be affected by different environmental conditions such as the temperature, pH, concentration of the chemical, amount and type of microorganisms, time, presence of other chemicals, and content of dissolved oxygen.

The Organization for Economic Co-operation and Development (OECD) distinguishes six forms of biodegradation classification in their guidelines:[15]

(a) ultimate biodegradation (mineralization), where the compound is totally used by microorganisms resulting in the production of biomass (new microbial cellular content), water, mineral salts, and carbon dioxide;

(b) primary biodegradation (biotransformation), where the chemical is transformed by biological action such that it will lose some specific property;

(c) readily biodegradable, are chemicals that will rapidly and readily biodegrade in an aquatic environment under aerobic conditions;

**Figure 3.7** Biodegradation in nature.[14]

(d) inherently biodegradable, are chemicals for which there is unequivocal evidence of biodegradation;

(e) half-life ($t_{0.5}$), which is the time to transform 50% of the substance, assuming first-order kinetics for the transformation reaction; and

(f) disappearance time 50 ($DT_{50}$), which is the time within which the initial concentration of the substance is reduced by 50%.

## 3.6 What Is Aerobic Degradation?

Aerobic degradation represents the breakdown of chemicals by microorganisms in the presence of oxygen (Fig. 3.7). Many organic contaminants are rapidly degraded under aerobic conditions. Aerobic bacteria use oxygen as a preferred electron acceptor to oxidize substrates in order to obtain energy. This process is initiated by the oxygenase and peroxygenase enzymes (Fig. 3.8). Following this initiation, peripheral metabolic pathways degrade the chemicals so that they can be used for the energy needs and growth of the microorganisms. Large numbers of natural and xenobiotic chemicals are used as a carbon source and electron donors and electron acceptors (such as sulfur, metal ions, methane, hydrogen) for the generation of energy by various bacterial species. Mixtures of different bacterial communities have the best potential for biodegradation since no single species has the enzymes to degrade all organic compounds. Figure 3.9 shows aerobic degradation pathways for alkanes and aromatics. Besides the variability of the microbial community, the rate and extent of degradation depends on the different environmental factors such as the oxygenation level, pH, temperature, and availability of nitrogen and phosphorus sources.

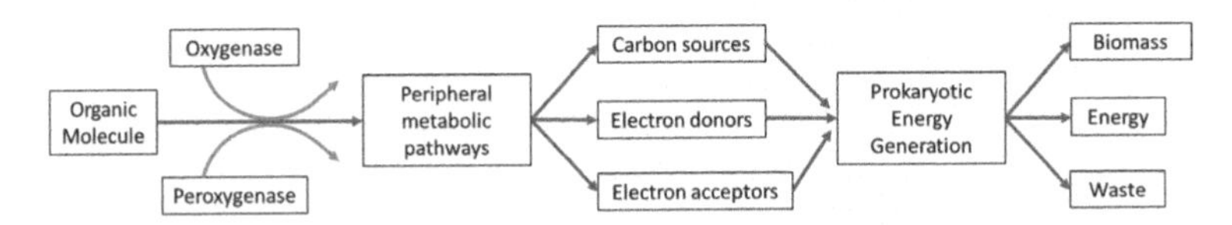


**Figure 3.8** Schematic representation of aerobic degradation.

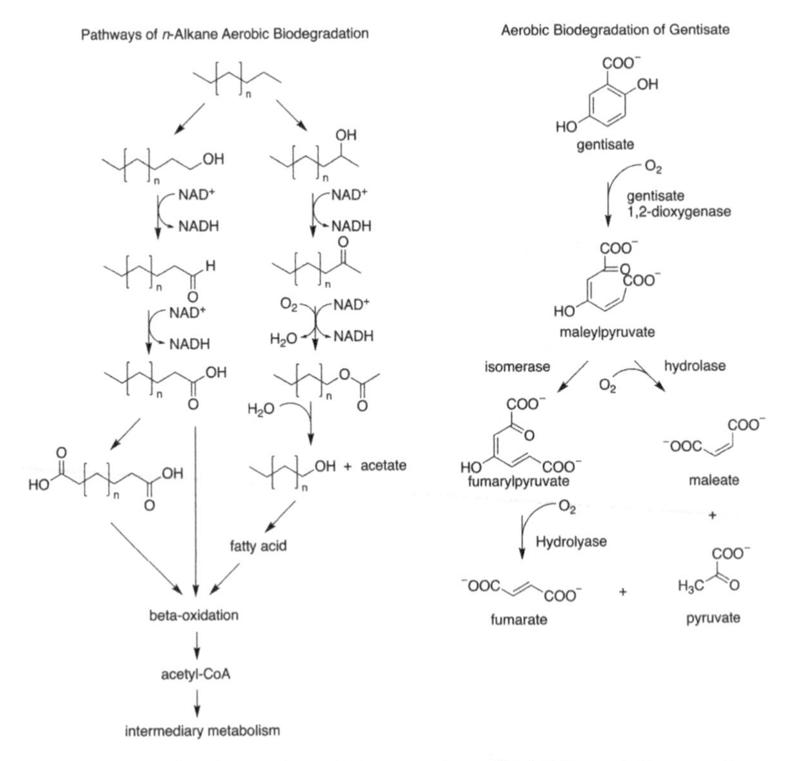


**Figure 3.9** (Left) Alkanes degradation reactions; (Right) Degradation reactions of aromatic molecule gentisate.[16,18]

Some of the bacteria that possess the highest potential for biodegradation are gram-negative *Pseudomonas*, where some species can use more than 100 different organic compounds as a carbon source.[16] Another important group is the gram-positive coryneform and *Rhodococcus* bacteria. *Rhodococcus* are particularly interesting since some groups of these bacteria possess lipophilic cell structures on the surface of the cell that is probably important for the affinity towards lipophilic pollutants.[17] Besides the direct use as a source of carbon and energy for the cells, microorganisms can be involved in co-metabolic biotransformation processes oxidizing chemicals without apparent gain. These products can then be used as products in the metabolism of other microorganisms present in the environment.

Microfungi, i.e., yeasts and molds, are capable of biodegrading both aliphatic and aromatic compounds.[18] Certain higher fungi have evolved oxidative systems for the degradation of lignin that also act as a co-metabolic degradation of some persistent organic pollutants.[19]

## 3.7 What Is Anaerobic Degradation?

In the environments where oxygen is not present, biodegradation of chemical compounds is performed by anaerobic microorganisms.[20] Anaerobic degradation often results in the formation of methane and $CO_2$ as final products. Aerobic bacteria generally obtain more energy than their anaerobic counterparts for the same substrate turnover. This means that anaerobic bacteria will produce less biomass, slowing their growth. While it is widely understood that anaerobic degradation is slower and less efficient than aerobic degradation, some chemical reactions occurring in anoxic environments can benefit more from the lack of oxygen. For example, the degradation of cellulose in the cow's rumen is much faster without oxygen.[21]

Degradation of organic molecules in the absence of oxygen seems to depend on the redox potential of alternative electron acceptors such as nitrate, manganese(IV) oxide, iron(III) hydroxides, sulfate, and $CO_2$. The degradation of molecules using these electron acceptors results in the release of nitrite, ammonia, $N_2$, manganese(II) and iron (II) carbonates, sulfides, and methane.[20d]

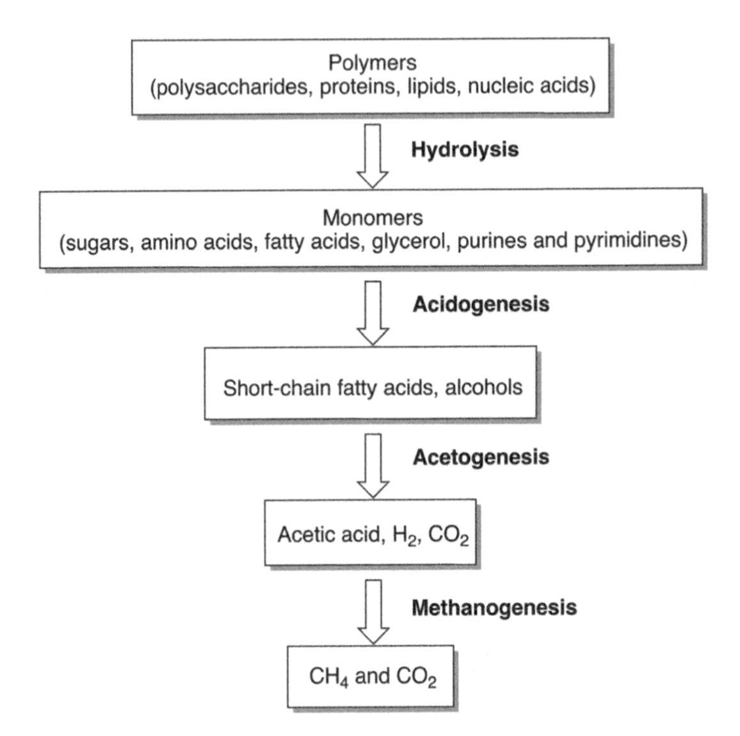


**Figure 3.10** Anaerobic digestion reactions.[20c]

Anaerobic degradation consists of four key biological and chemical stages: hydrolysis, acidogenesis, acetogenesis, and methanogenesis (Fig. 3.10). Usually, anaerobic biodegradation of organic compounds starts with hydrolysis because most biomass undergoing decomposition consists of large organic polymers. Hydrolysis breaks these polymers into smaller fragments, i.e., amino acids, fatty acids, and simple sugars. This first stage can release acetate and hydrogen that can be directly used in methanogenesis, while other generated compounds need to be transformed, i.e., catabolized, into other molecules that can undergo methanogenesis. The second stage, i.e., acidogenesis, is a process in which fermentative bacteria transform products of the previous phase into carbonic acids, alcohols, ammonia, $CO_2$, and $H_2S$. In the third phase, i.e., acetogenesis, simple molecules generated through acidogenesis are mostly transformed into acetic acid, $CO_2$, and $H_2$. In the terminal stage of anaerobic biodegradation, i.e., methanogenesis, intermediate products from the previous phases are converted into methane, $CO_2$, and water.

A broad range of chemical compounds can be biodegraded anaerobically: aliphatic hydrocarbons are usually degraded if they contain unsaturated bonds; ethers can be degraded if they can be transformed into hemiacetals; ketones' primary activation reaction is carboxylation; mononuclear aromatic compounds are efficiently degraded if they contain at least one amino, carboxy, hydroxy, methoxy, or methyl substituent; halogenated aliphatic and aromatic compounds are reductively dehalogenated; sulfonates are not as effectively degraded; nitro- and azo-substituted compounds are effectively degraded through reduction (Fig. 3.11).

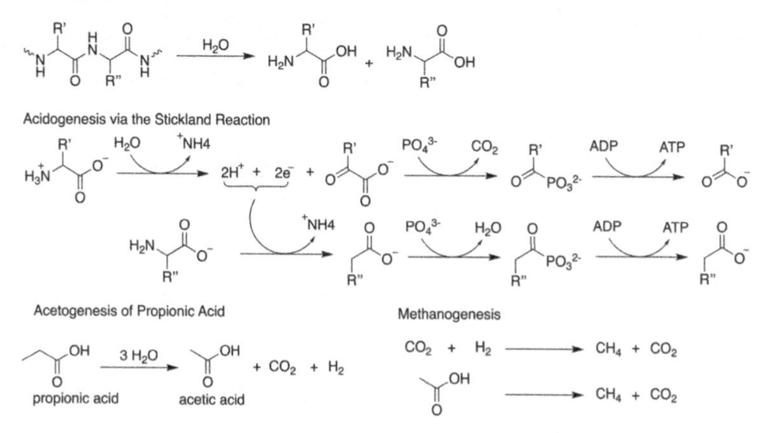


**Figure 3.11** Example of the anaerobic reactions indicated in Fig 3.10.[22]

Typical bacteria genera involved in the anaerobic biodegradations are *Clostridia*, *Eubacteria*, *Syntrophobacteria*, and *Methanosaecenia*.

## 3.8 How Does Branching of a Carbon Chain Impact Biodegradation?

Branching generally reduces the rate of biodegradation, mostly with the complex branching chains such as the *t*-butyl group, while simpler methyl branching does not impact the biodegradation rate in a significant manner. For example, tetrapropylene alkylbenzene sulfonate is a highly branched molecule that was degraded only by about 50% in a sewage treatment process.[23] Of course, there are exemptions to this rule, e.g., cholesterol, vitamin A, pantothenic acid, and pentaerythritol.[1] Additionally, it was found that the extensive methyl branching in the alkyl chain leads to reduced biodegradation rates. This is probably because these hydrocarbons are not easily absorbed into the cells, or they are not susceptible to oxidation enzymes.[24] It was shown that synthetic-based fluids follow the rules of thumb stating that linear hydrocarbons and esters degrade much faster than branched and cyclic compounds and diesel, which is a mixture of aliphatic, cyclic, branched, and unbranched compounds.[25] Also, unsaturated hydrocarbons biodegrade much faster than their saturated counterparts. Figure 3.12 shows several examples that demonstrate the influence of branching on biodegradation.

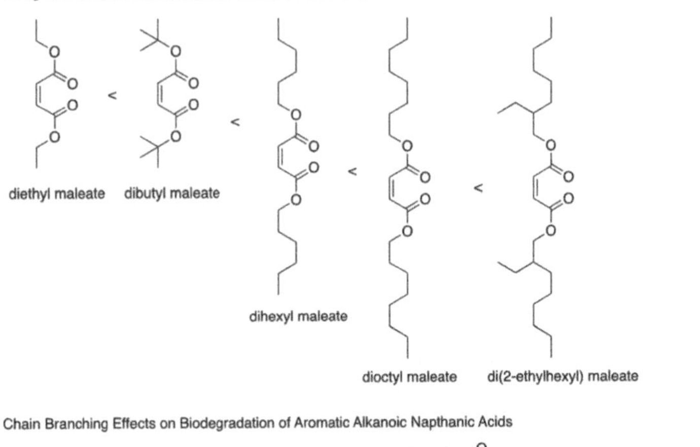


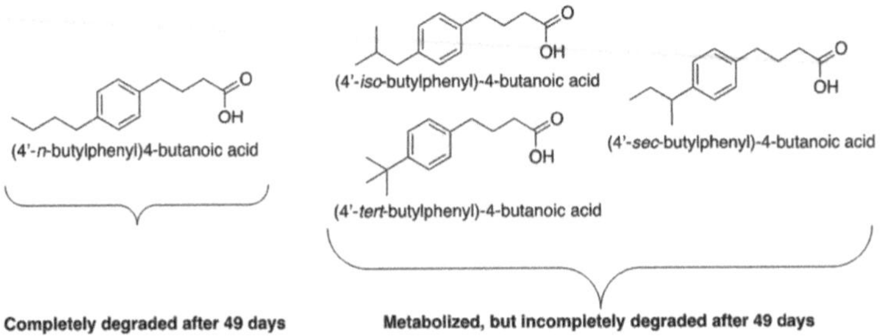


**Figure 3.12** Selected examples demonstrating increased biodegradation due to branching.[26]

## 3.9 How Do Fused Rings Impact Biodegradation?

Polycyclic aromatic hydrocarbons (PAHs) are a large group of environmental pollutants, mostly originating from the incomplete combustion of various organic compounds. Generally, once these substances get into the environment, microbial degradation is the most efficient way for their degradation. The PAHs molecular structure consists of two or more fused six-carbon aromatic rings, and this property generally determines their rate of biodegradation. Usually, lower molecular weight PAHs, i.e., those with fewer fused rings, are biodegraded at a faster rate than those with higher molecular weight.[27] Figure 3.13 shows the potential degradation of naphthalene by aerobic bacteria. The high-molecular weight compounds, with four or more fused rings, are less water-soluble, more lipophilic, and less volatile than those with lower molecular weight. However, it was noticed that the simultaneous presence of multiple PAHs in the same environment can affect their biodegradation rates, e.g., when acenaphthene, anthracene,

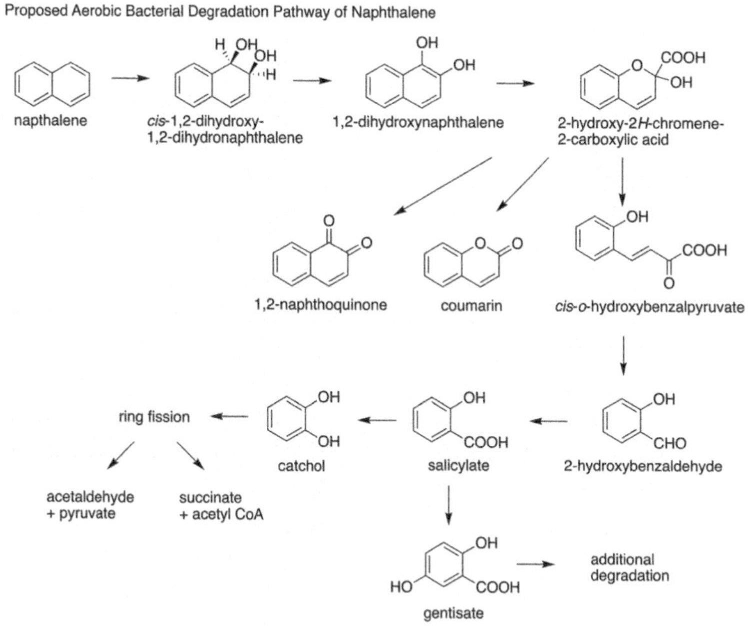

Proposed Aerobic Bacterial Degradation Pathway of Naphthalene

napthalene → *cis*-1,2-dihydroxy-1,2-dihydronaphthalene → 1,2-dihydroxynaphthalene → 2-hydroxy-2*H*-chromene-2-carboxylic acid

1,2-naphthoquinone; coumarin; *cis*-*o*-hydroxybenzalpyruvate

ring fission ← catechol ← salicylate ← 2-hydroxybenzaldehyde

acetaldehyde + pyruvate; succinate + acetyl CoA

gentisate → additional degradation

**Figure 3.13** Proposed aerobic bacterial degradation pathway of naphthalene.[29]

fluorene, phenanthrene, pyrene, and benzo[a]pyrene are present in the mixture, biodegradation is reduced for phenanthrene and acenaphthene, and enhanced for the others, except for benzo[a] pyrene which was not biodegraded.[28]

## 3.10 How Do Heteroatoms Impact Biodegradation?

Studies of anaerobic biodegradation of compounds containing nitrogen, oxygen, and sulfur heterocyclic compounds show that the nitrogen and oxygen-containing compounds were more susceptible than those containing sulfur atoms. Additionally, compounds containing carboxy substituents were more readily metabolized by anaerobic microbes than unsubstituted, or methylated, compounds.[30] The presence of C-F bonds, tertiary and quaternary N atoms, and ether groups result in reduced biodegradability.[31] However, if the oxygen is present in the form of esters or alcohols, biodegradability will be increased.[32] The study of biodegradation of plant biomass has shown that the materials with S- and P-containing molecules had a much stronger effect than the material with N-containing molecules.[33] S- and P-containing molecules made crop residues more available for microbial biodegradation, leading to higher biodegradation rates. The influence of heteroatom presence in the molecule on its biodegradation rate is shown in Fig. 3.14.

Generally, it was observed that polymers with heteroatoms, such as polyamines or polyesters, show higher degradation rates than those containing only carbon backbones. Hetero-groups containing oxygen promote the formation of carbanions in the presence of a base by influencing neighboring C-H bonds. This makes polymers containing these groups more susceptible to thermal degradation and biodegradation. However, in the case of polyesters containing aromatic groups, even the presence of the ester bond does not enable biodegradation.[34] The study of polymer degradation in the marine environment indicates that plastics containing heteroatoms in the main polymer chain can be degraded by hydrolysis, photooxidation, and biodegradation. PET is highly resistant to biodegradation because of the very compact structure due to the presence of aromatic groups, while polyurethane is much more susceptible to fungal biodegradation by cleaving the ester bond of polyurethane.[35]

| | phosphate ester | carboxylic acid | ester | amide | aldehyde | aliphatic alcohol | aliphatic amine | aliphatic Cl | aromatic nitro | aromatic Cl | trifluoromethyl | tertiary amine |
|---|---|---|---|---|---|---|---|---|---|---|---|---|
| **primary biodegradation coefficient** | 0.465 | 0.386 | 0.229 | 0.205 | 0.197 | 0.129 | 0.043 | -0.101 | -0.108 | -0.165 | -0.274 | -0.288 |

**Figure 3.14** Influence of the heteroatoms on biodegradation rates.[31]

## 3.11 How Do Charged Functional Groups Impact Biodegradation?

Class-specific QSAR studies show that charged chemicals are inherently less biodegradable, with this effect being more pronounced for the molecules with high $M_w$.[36] Charge distribution over the neutral molecules can affect their biodegradation rate by influencing the binding to specialized enzymes or the cell membrane permeability of the microorganisms doing the degradation.

A limited study of 18 micropollutant compounds reported that all negatively charged molecules were characterized by high biodegradation rates, while positively charged molecules exhibited lower biodegradation rates.[37]

## 3.12 What Is Persistence?

The environmental fate of large numbers of chemicals is decomposition, i.e., degradation either abiotically, or by the work of microorganisms. However, some molecules (especially synthetic human-made ones) can resist this degradation, thus they persist in the environment and can bioaccumulate causing various adverse effects as mentioned earlier (Fig. 3.15).

The International Council of Chemical Associations (ICCA) describes persistence as the ability of a chemical to stay unchanged in the environment for a long period of time.[38] This persistence can result in environmental issues because the removal of the chemical from the environment can be very slow, resulting in accumulations that can cause hazards, allow for long-distance transportation to areas far removed from the exposure site, and the concentration of chemical may take a long time to go back to no-effect levels. The European Centre for Ecotoxicology and Toxicology of Chemicals (ECETOC) suggests that persistent substances should be those resistant to abiotic and/or biotic degradation under both aerobic and anaerobic conditions.[38]

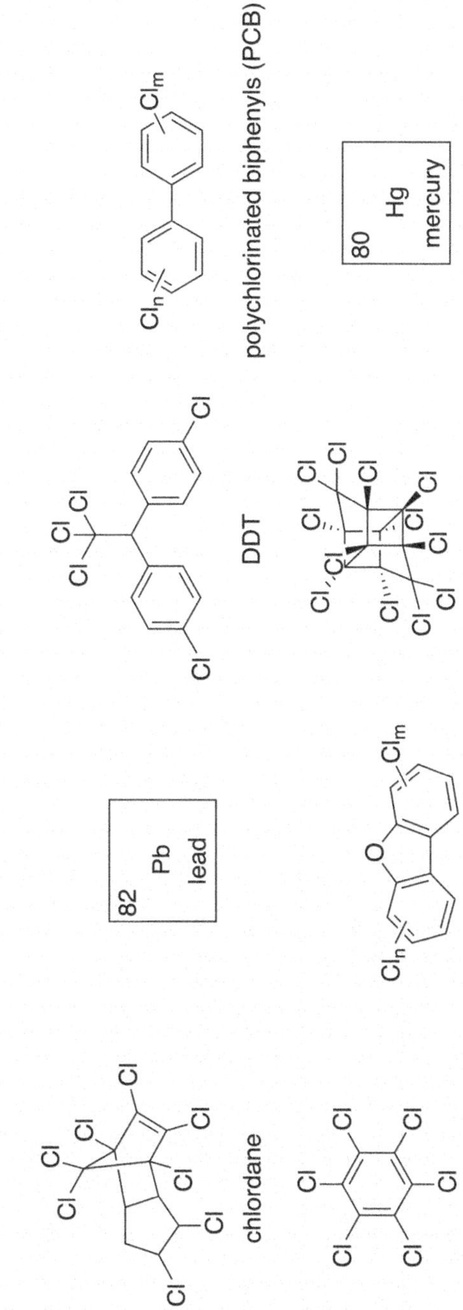


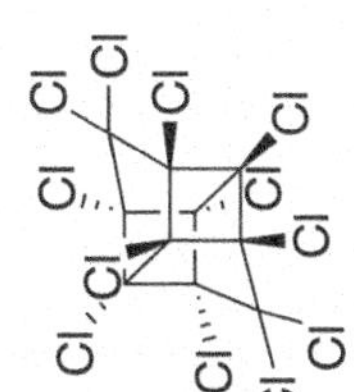


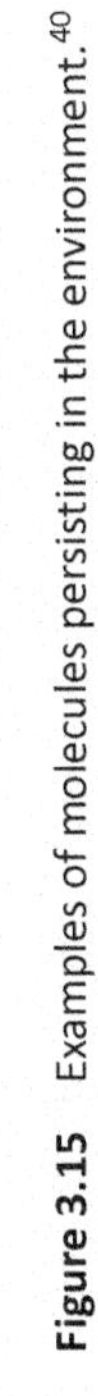

**Figure 3.15** Examples of molecules persisting in the environment.[40]

The persistence of chemical compounds cannot be measured directly, so it is often estimated based on the measurements of degradation in the laboratory, or by the observation of the continuing presence of the chemical in the environment.[38,39] Regulatory agencies define persistence based on the half-lives of chemicals; however, this is not the best method because this assumes that all chemicals decompose following the first-rate kinetics. Additionally, it does not take into consideration the environmental conditions that affect the degradation at the site of exposure, the possible transport of the chemicals, and the toxicity of the chemicals. Thus, a substance's half-life is estimated and defined for different environmental compartments, i.e., air, fresh water, seawater, soil, and sediment.

## 3.13 Why Are Perfluorinated Compounds so Persistent?

Perfluorinated compounds (PFCs) are organic molecules in which all of the hydrogens from the hydrocarbon backbone are replaced by fluorine atoms. Because they possess extremely stable C-F bonds, these molecules have very high chemical and thermal stability, which made them very interesting and useful for a lot of applications (Fig. 3.16).[41] They have been extensively used for the production of surfactants, or polymers as grease and soil repellants in the textile and food industry. However, the same stability makes them very resistant to the degradation processes, so they persist in the environment and bioaccumulate causing further adverse effects, e.g., accumulation in the tissues and nephrotoxicity.[42]

PFCs are a large group of substances that can be divided into perfluorinated sulfonic acids, perfluorinated carboxylic acids (PFCAs), low- and high-molecular-weight fluoropolymers, and fluorotelomer alcohols. Of these, perfluorooctanesulfonic acid (PFOS) and perfluorooctanoic acid (PFOA) are the most studied from a toxicological perspective (Fig. 3.17).

PFOA and PFOS are very water-soluble, weakly lipophilic, and bind preferentially to proteins such as albumin, b-lipoproteins, or fatty acid-binding proteins in the liver.[44] The binding strength of the PCAs to bovine serum albumin increases with the longer chain compounds mostly because of the increase in non-covalent binding capacity influenced by the van der Waals force and hydrogen bonds.[41] According to current scientific knowledge, PFOA and PFOS are not metabolized in mammals so they do not get defluorinated and do not enter phase II in metabolism.[44] Because they cannot be biotransformed, the only way to remove them from the mammal body is by excretion via specialized transport proteins. For example, PFC half-life values based on the serum concentrations taken from the workers exposed in the fluorochemical industry were quite long, with mean times being 3.8 years for PFOA, 5.4 years for PFOS, and 8.5 years for perfluorohexanesulfonic acid (PFHxS).[45]

**Figure 3.16** Real-life products based on PFCs. (a) Nonstick pan—teflon®; (b) Flame-retardant foams; (c) Hydrophobic materials—Gore-Tex®.[43]

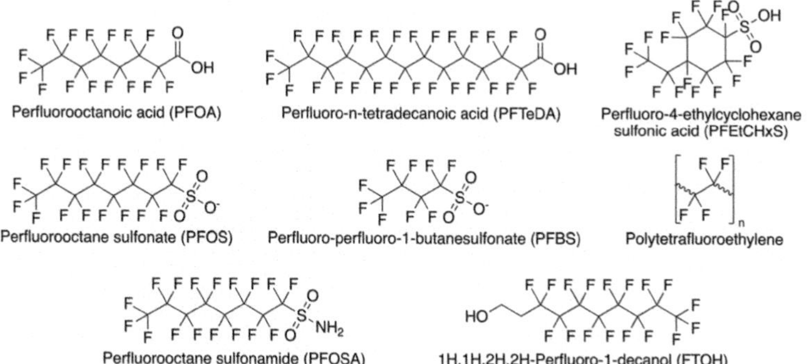


**Figure 3.17** Structures of selected perfluorinated compounds.

## References


1. Boethling, R. S.; Sommer, E.; DiFiore, D. Designing small molecules for biodegradability. *Chem Rev*, 2007, **107**(6), 2207–2227.
2. Speight, J. G. Redox transformations. In: Speight, J. G. (Ed.) *Reaction Mechanisms in Environmental Engineering*, Elsevier, 2018, pp. 231–267.
3. Karlapudi, A. P.; Venkateswarulu, T. C.; Tammineedi, J.; Kanumuri, L.; et al. Role of biosurfactants in bioremediation of oil pollution: a review. *Petroleum*, 2018, **4**(3), 241–249.
4. (a) Albini, A.; Germani, L. Chapter 1: Photochemical methods. In: Albini, A.; Fagnoni, M. (Eds.), *Handbook of Synthetic Photochemistry*, Wiley-VCH: Weinheim, Germany, 2009, pp. 1–24; (b) Bochet, C. G.; Blanc, A. Photolabile protecting groups in organic synthesis. In: Albini, A.; Fagnoni, M. (Eds.), *Handbook of Synthetic Photochemistry*, Wiley-VCH: Weinheim, Germany, 2010, pp. 417–447; (c) Portmann, R. W.; Daniel, J. S.; Ravishankara, A. R. Stratospheric ozone depletion due to nitrous oxide: influences of other gases. *Philos Trans R Soc Lond B, Biol Sci*, 2012, **367**(1593), 1256–1264; (d) Yousif, E.; Haddad, R. Photodegradation and photostabilization of polymers, especially polystyrene: review. *Springerplus*, 2013, **2**(1), 398.
5. Shams, M.; Guiney, L. M.; Huang, L.; Ramesh, M.; et al. Influence of functional groups on the degradation of graphene oxide nanomaterials. *Environ Sci Nano*, 2019, **6**(7), 2203–2214.
6. Nawaz, T.; Sengupta, S. Chapter 4: Contaminants of emerging concern: occurrence, fate, and remediation. In: Ahuja, S. (Ed.), *Advances in Water Purification Techniques*, Elsevier, 2019, pp. 67–114.
7. Mill, T. Hydrolysis and oxidation processes in the environment. *Environ Toxicol Chem*, 1982, **1**(2), 135–141.
8. McMurry, J. *Organic Chemistry*. Thomson Brooks/Cole: Belmont, 2008.
9. Shen, M.; Almallahi, R.; Rizvi, Z.; Gonzalez-Martinez, E.; et al. Accelerated hydrolytic degradation of ester-containing biobased epoxy resins. *Polym. Chem.*, 2019, **10**(23), 3217–3229.
10. (a) Tomczyk, A.; Sokołowska, Z.; Boguta, P. Biochar physicochemical properties: pyrolysis temperature and feedstock kind effects. *Rev Environ Sci Biotechnol*, 2020, **19**(1), 191–215; (b) Zaman, C. Z.; Pal, K.; Yehye, W. A.; Sagadevan, S.; et al. Chapter 1: Pyrolysis: a sustainable way to generate energy from waste. In: Samer, M. (Ed.), *Pyrolysis*, IntechOpen, 2017.

11. Beyler, C. L.; Hirschler, M. M. Thermal decomposition of polymers. In: DiNenno, P. J.; Drysdale, D. (Eds.), *SFPE Handbook of Fire Protection Engineering*, 3rd ed., National Fire Protection Association: Quincy, MA, 2002, pp. 100–132.

12. (a) Shen, D.; Xiao, R.; Gu, S.; Zhang, H. The overview of thermal decomposition of cellulose in lignocellulosic biomass. In: Kadla, J.; Van de Ven, T. G. M. (Eds.), *Cellulose: Biomass Conversion*, IntechOpen, 2013; (b) Shen, D. K.; Gu, S. The mechanism for thermal decomposition of cellulose and its main products. *Bioresour Technol*, 2009, **100**(24), 6496–6504.

13. (a) Means, J. L.; Anderson, S. J. Comparison of five different methods for measuring biodegradability in aqueous environments. *Water Air Soil Pollut*, 1981, **16**(3), 301–315; (b) Yabannavar, A. V.; Bartha, R. Methods for assessment of biodegradability of plastic films in soil. *Appl Environ Microbiol*, 1994, **60**(10), 3608–3614; (c) Castellani, F.; Esposito, A.; Stanzione, V.; Altieri, R. Measuring the biodegradability of plastic polymers in olive-mill waste compost with an experimental apparatus. *Adv Mater Sci Eng*, 2016, **2016**, 1–7; (d) Speight, J. Removal of organic compounds from the environment. *Environmental Organic Chemistry for Engineers*, Elsevier: Kidlington, Oxford, 2017, pp. 387–432.

14. Ito, J. Neighbor's compost (CC BY 2.0). https://www.flickr.com/photos/35034362831@N01/423508812 (accessed December 1).

15. European Centre for Ecotoxicology and Toxicology of Chemicals. Definition(s) according to OECD. https://www.ecetoc.org/report/measured-partitioning-property-data/biodegradation/definitions-according-to-oecd/ (accessed September 1).

16. Fritsche, W.; Hofrichter, M. Aerobic degradation of recalcitrant organic compounds by microorganisms. In: Jördening, H. J.; Winter, J. (Eds.) *Environmental Biotechnology*, Wiley-VCH: Germany, 2004, pp. 203–227.

17. Mohapatra, P. K. *Textbook of Environmental Microbiology*. I. K. International Publishing House Pvt. Ltd.: New Delhi, India, 2008.

18. Pérez-Pantoja, D.; González, B.; Pieper, D. Aerobic degradation of aromatic hydrocarbons. In: Timmis, K. N. (Ed.), *Handbook of Hydrocarbon and Lipid Microbiology*, Vol. 2, Springer-Verlag: Berlin, 2010, pp. 799–837.

19. Deshmukh, R.; Khardenavis, A. A.; Purohit, H. J. Diverse metabolic capacities of fungi for bioremediation. *Indian J Microbiol*, 2016, **56**(3), 247–264.

20. (a) Meegoda, J. N.; Li, B.; Patel, K.; Wang, L. B. A review of the processes, parameters, and optimization of anaerobic digestion. *Int J Environ Res Public Health*, 2018, **15**(2224), 16; (b) Rabus, R.; Boll, M.; Heider, J.; Meckenstock, R. U.; et al. Anaerobic microbial degradation of hydrocarbons: from enzymatic reactions to the environment. *Microb Physiol*, 2016, **26**(1–3), 5–28; (c) Sikora, A.; Detman, A.; Chojnacka, A.; Blaszczyk, M. Anaerobic digestion: I. A common process ensuring energy flow and the circulation of matter in ecosystems. II. A tool for the production of gaseous biofuels. In: Jozala, A. (Ed.), *Fermentation Processes*, IntechOpen, 2017; (d) Ghattas, A.-K.; Fischer, F.; Wick, A.; Ternes, T. A. Anaerobic biodegradation of (emerging) organic contaminants in the aquatic environment. *Water Res*, 2017, **116**, 268–295.

21. Zhang, L.; Chung, J.; Jiang, Q.; Sun, R.; et al. Characteristics of rumen microorganisms involved in anaerobic degradation of cellulose at various pH values. *RSC Adv*, 2017, **7**(64), 40303–40310.

22. (a) Angelidaki, I.; Karakashev, D.; Batstone, D. J.; Plugge, C. M.; et al. Biomethanation and its potential. In: Rosenzweig, A. C.; Ragsdale, S. W. (Eds.), *Methods in Enzymology*, Vol. 494, Academic Press, 2011, pp. 327–351; (b) de Vladar, H. P. Amino acid fermentation at the origin of the genetic code. *Biol Direct*, 2012, **7**(1), 6; (c) Murali, N.; Srinivas, K.; Ahring, B. K. Biochemical production and separation of carboxylic acids for biorefinery applications. *Fermentation*, 2017, **3**(22), 25.

23. Painter, H. A. Anionic surfactants. In: de Oude, N. T. (Ed.), *Detergents. Anthropogenic Compounds*, Vol. 3/3F, Springer: Berlin, Heidelberg, 1992, pp. 1–88.

24. Berekaa, M. M.; Steinbüchel, A. Microbial degradation of the multiply branched alkane 2,6,10,15,19,23-hexamethyltetracosane (squalane) by *Mycobacterium fortuitum* and *Mycobacterium ratisbonense*. *Appl Environ Microbiol*, 2000, **66**(10), 4462–4467.

25. (a) Sugiura, K.; Ishihara, M.; Shimauchi, T.; Harayama, S. Physicochemical properties and biodegradability of crude oil. *Environ Sci Technol*, 1997, **31**(1), 45–51; (b) Norman, M., Ross, S., McEwen, G.; Getliff, J. *Minimizing Environmental Impacts and Maximizing Hole Stability: Significance of Drilling with Synthetic Fluids in NZ*, New Zealand Petroleum Conference, Carlton Hotel, Auckland, New Zealand, February 24–27, 2002.

26. (a) Johnson, R. J.; Smith, B. E.; Sutton, P. A.; McGenity, T. J.; et al. Microbial biodegradation of aromatic alkanoic naphthenic acids is affected by the degree of alkyl side chain branching. *ISME J*, 2011, **5**(3), 486–496; (b) Erythropel, H. C.; Brown, T.; Maric, M.; Nicell, J. A.; et al. Designing

greener plasticizers: effects of alkyl chain length and branching on the biodegradation of maleate based plasticizers. *Chemosphere*, 2015, **134**, 106–112.

27. (a) Park, K. S.; Sims, R. C.; Dupont, R. R. Transformation of PAHs in soil systems. *J Environ Eng*, 1990, **116**(3), 632–640; (b) Hassanshahian, M.; Abarian, M.; Cappello, S. Biodegradation of aromatic compounds. In: Chamy, R.; Rosenkranz, F. (Eds.), *Biodegradation and Bioremediation of Polluted Systems*, IntechOpen, 2015.
28. Yuan, S. Y.; Chang, J. S.; Yen, J. H.; Chang, B.-V. Biodegradation of phenanthrene in river sediment. *Chemosphere*, 2001, **43**(3), 273–278.
29. Seo, J.-S.; Keum, Y.-S.; Li, Q. X. Bacterial degradation of aromatic compounds. *Int J Environ Res Public Health*, 2009, **6**(1), 278–309.
30. Kuhn, E. P.; Suflita, J. M. Microbial degradation of nitrogen, oxygen and sulfur heterocyclic compounds under anaerobic conditions: studies with aquifer samples. *Environ Toxicol Chem*, 1989, **8**(12), 1149–1158.
31. Boethling, R. S.; Howard, P. H.; Meylan, W.; Stiteler, W.; et al. Group contribution method for predicting probability and rate of aerobic biodegradation. *Environ Sci Technol*, 1994, **28**(3), 459–465.
32. Kümmerer, K. Rational design of molecules by life cycle engineering. In: Kümmerer, K.; Hempel, M. (Eds.), *Green and Sustainable Pharmacy*, Springer: Heidelberg, 2010, pp. 135–146.
33. Caricasole, P.; Hatcher, P. G.; Ohno, T. Biodegradation of crop residue-derived organic matter is influenced by its heteroatomic stoichiometry and molecular composition. *Appl Soil Ecol*, 2018, **130**, 21–25.
34. Sheel, A.; Pant, D. Microbial depolymerization. In: Varjani, S. J.; Gnansounou, E.; Gurunathan, B.; Pant, D. (Eds.), *Waste Bioremediation*, Springer: Singapore, 2018, pp. 61–103.
35. Gewert, B.; Plassmann, M. M.; MacLeod, M. Pathways for degradation of plastic polymers floating in the marine environment. *Environ Sci Process Impacts*, 2015, **17**(9), 1513–1521.
36. Nolte, T. M.; Pinto-Gil, K.; Hendriks, A. J.; Ragas, A. M. J.; et al. Quantitative structure–activity relationships for primary aerobic biodegradation of organic chemicals in pristine surface waters: starting points for predicting biodegradation under acclimatization. *Environ Sci Process Impacts*, 2018, **20**(1), 157–170.
37. Bertelkamp, C.; Reungoat, J.; Cornelissen, E. R.; Singhal, N.; et al. Sorption and biodegradation of organic micropollutants during river bank filtration: a laboratory column study. *Water Res*, 2014, **52**, 231–241.

38. European Centre for Exotoxicology and Toxicology of Chemicals. *Persistence of Chemicals in the Environment*; ECETOC Technical ReportEuropean Centre for Exotoxicology and Toxicology of Chemicals: Brussels, Belgium, 2003, p. 199.
39. Gouin, T.; Mackay, D.; Webster, E.; Wania, F. Screening Chemicals for Persistence in the Environment. *Environmental Science & Technology* 2000, **34**(5), 881–884.
40. UN Envrionment Programme. All POPs listed in the Stockholm Convention. http://chm.pops.int/TheConvention/ThePOPs/ListingofPOPs/tabid/2509/Default.aspx (accessed Feb 17).
41. Qin, P.; Liu, R.; Pan, X.; Fang, X.; et al. Impact of carbon chain length on binding of perfluoroalkyl acids to bovine serum albumin determined by spectroscopic methods. *J Agric Food Chem*, 2010, **58**(9), 5561–5567.
42. (a) Stahl, T.; Mattern, D.; Brunn, H. Toxicology of perfluorinated compounds. *Environ Sci Eur*, 2011, **23**(1), 52; (b) Ferrari, F.; Orlando, A.; Ricci, Z.; Ronco, C. Persistent pollutants: focus on perfluorinated compounds and kidney. *Curr Opin Crit Care*, 2019, **25**(6), 539–549.
43. (a) National Science Foundation. Super-hydrophobic coating (public domain). https://commons.wikimedia.org/wiki/File:Super-hydrophobic_Coating.jpg (accessed May 1); (b) trozzolo. Happy Pan Poster (public domain). https://commons.wikimedia.org/wiki/File:Happy_Pan_Poster.jpg (accessed May 1); (c) Tello, J. Waterproof (CC BY 2.0). https://www.flickr.com/photos/17333088@N00/2938584806 (accessed May 1).
44. Kudo, N.; Kawashima, Y. Toxicity and toxicokinetics of perfluorooctanoic acid in humans and animals. *J Toxicol Sci*, 2003, **28**(2), 49–57.
45. Olsen, G. W.; Burris, J. M.; Ehresman, D. J.; Froehlich, J. W.; et al. Half-life of serum elimination of perfluorooctanesulfonate,perfluorohexanesulfonate, and perfluorooctanoate in retired fluorochemical production workers. *Environ Health Perspect*, 2007, **115**(9), 1298–1305.

Chapter 4

# Dose/Response/Risk

## 4.1 What Is a Dose?

As mentioned earlier in the text, when the xenobiotic enters the body, its behavior will in part depend on the amount that was absorbed or administered, i.e., its dose. In other words, a dose can be considered the amount of a substance administered (as a medication) over some period of time or a measurement of environmental exposure. Toxicologists usually distinguish between the individual doses and total dose, expressed as the sum of all individual doses.[1] Individual dose is the administered dose, i.e., the quantity that is administered orally or by injection—more relevant in medicine—and absorbed (internal) dose, i.e., the amount of substance that entered the body through the exposure routes. This differentiation both describes the amount of the substance that enters the body and more importantly the amount that is actually absorbed in the body. The absorbed dose of a substance is usually expressed as a milligram of the substance per kilogram of body weight, i.e., mg/kg.

A famous myth of king Mithridates VI of Pontus (Fig. 4.1) tells a story of the poison dosage.[3] According to the myth and the writings of Pliny the Elder (an ancient Roman philosopher), the king was afraid that he would be assassinated by some poison, so

*First Do No Harm: A Chemist's Guide to Molecular Design for Reduced Hazard*
Predrag V. Petrovic and Paul T. Anastas

ISBN 978-981-4968-59-1 (Hardcover), 978-1-003-35964-7 (eBook)
www.jennystanford.com

he consumed small amounts of a mixture of known poisons dried and prepared as a tablet to protect himself. While we cannot know for certain if this panacea worked, it illustrates the concept of dose fractionation. The total dose of the substance that the body absorbs does not necessarily mean its adverse effects will be expressed. If the individual dose is small enough, and the exposure is spread over a period of time that will allow for the substances to be metabolized, it will cause no harm. The strategy of dose fractionation is often used to improve the preclinical effectiveness of the drugs.[4]

**Figure 4.1** Portrait of the King of Pontus Mithridates VI as Heracles. Marble, Roman imperial period (1st century).[2]

## 4.2 What Is Dose-Response?

When considering the adverse effects of chemicals, toxicology relies on the concept of the dose-response relationship which connects the exposure to a chemical and changes in health.[5] This relationship is based on data gathered from experimental studies ranging from testing chemical compounds on cultured cell lines or animals to human clinical trials. The dose-response relationship provides several important bits of information, i.e., it establishes causality between a chemical and an observed effect, reveals the lowest dose where an effect is observed and gives the rate at which that effect is exhibited.[5b] However, before dose-response relationships can be used appropriately, several assumptions need to be considered. First, a cause-and-effect relationship between the response and the chemical needs to be established. Secondly, the magnitude of the response needs to be related to the dose, i.e., there has to exist a molecular target site (or sites) where the chemical may cause the response, which will depend on the concentration of the chemical, and accordingly, upon the administered dose. Finally, a method to quantifiably measure the dose-response relationship, and the precise way to exhibit the effect, need to exist. Often, chemicals can have multiple sites or mechanisms to express effects that each follow their own dose-response relationship. Additionally, it is important to emphasize that this approach cannot be applied to all xenobiotics. Notably, carcinogens and heavy metals do not follow the same rules as other chemicals.[6] The dose-response models for chemical carcinogens have changed with the advancement in the understanding of the toxic mechanism of action, where both proliferation and mutation of cells are followed separately. The last decade has seen important breakthroughs in transcriptomic analysis for the toxicological assessment of chemical carcinogens.[6b]

The practical considerations of dose-response relationships often distinguish between two types: the individual, or graded dose-response; and the quantal dose-response relationship (Fig. 4.2). The graded relationships describe the severity of the response in an individual in the population, which is dependent on the concentration of a chemical. On the other hand, quantal relationships indicate whether or not the response to the exposure exists across the population, i.e., the percentage of the population achieving the response at a given dose.

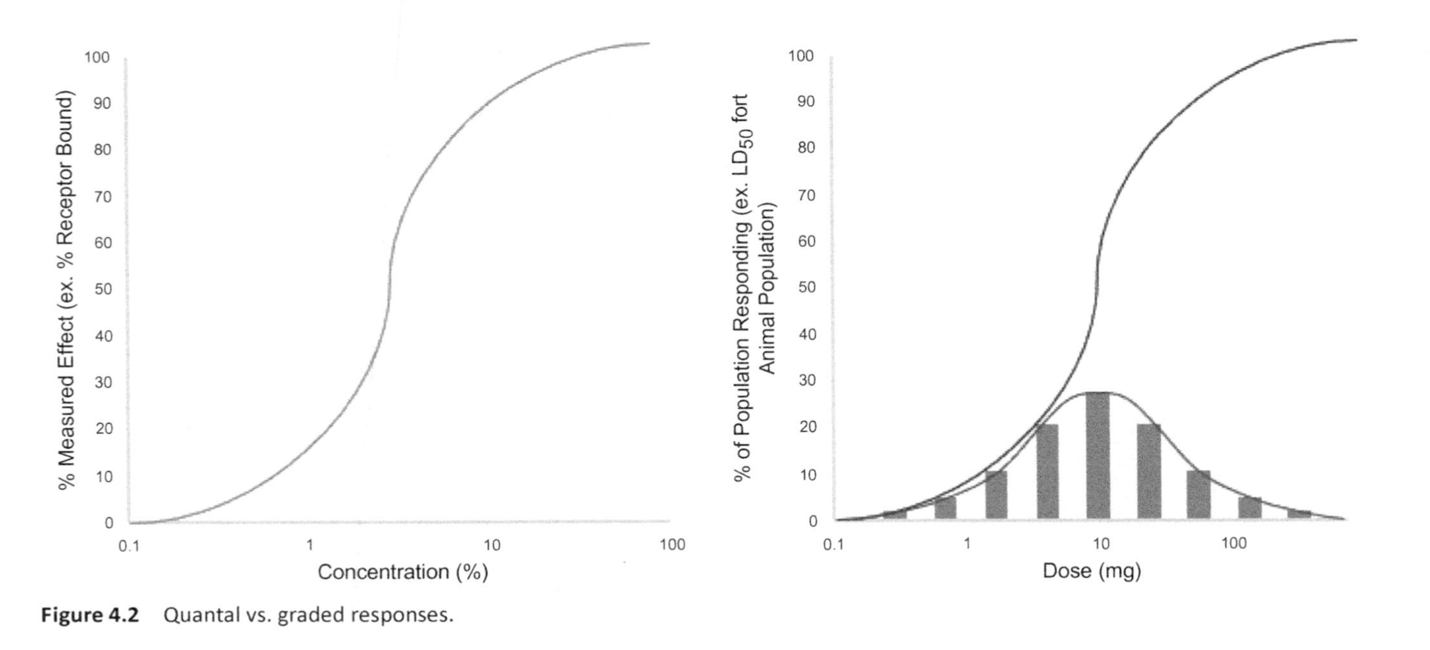


**Figure 4.2** Quantal vs. graded responses.

## 4.3 What Can Dose-Response Curves Tell Us?

The dose-response relationship can be graphically represented by the dose-response curve where the rate of response for a certain population is plotted ($y$-axis) in the range of doses of a substance ($x$-axis) (Fig. 4.3). When plotted on a linear scale, the resulting curve is usually hyperbolic. But typically, since the dose plotted on the $x$-axis is expressed on a base 10 of a logarithmic scale, the graph takes a sigmoidal ("S") shape. The logarithmic representation is often more useful, as it expands the dose scale in the region where the rapid change occurs, while the scale is compressed in the regions where large changes in dose do not have a great effect on the response. In other words, this representation better describes the potency of the chemical compound.

When we take a look at Fig. 4.3, there are different pieces of information that can be gathered:[5]

(a) The relationship reveals whether the exposure has caused an effect.

(b) The slope of the curve gives information on the rate at which the effect of the chemical is expressed. For example, a shallow slope indicates the weaker potency of a chemical, while a steeper slope indicates increased potency. Increased potency suggests a higher chance that adverse effects will be exhibited.

(c) The threshold dose, i.e., the distance from the graph's origin on the $x$-axis, represents the smallest dose for which the effect is observed, and it is an important indicator to consider when assessing the safe level of exposure to a substance.

## 4.4 What Is Toxic Dose Versus Effective Dose?

In toxicology and pharmacology, the relative safety of chemical substances is estimated using effective and toxic dose levels (Fig. 4.4).[5a] This relationship is often inverse for environmental pollutants and pharmaceuticals, i.e., the desired effect in a drug is the undesired effect with a chemical in the environment.

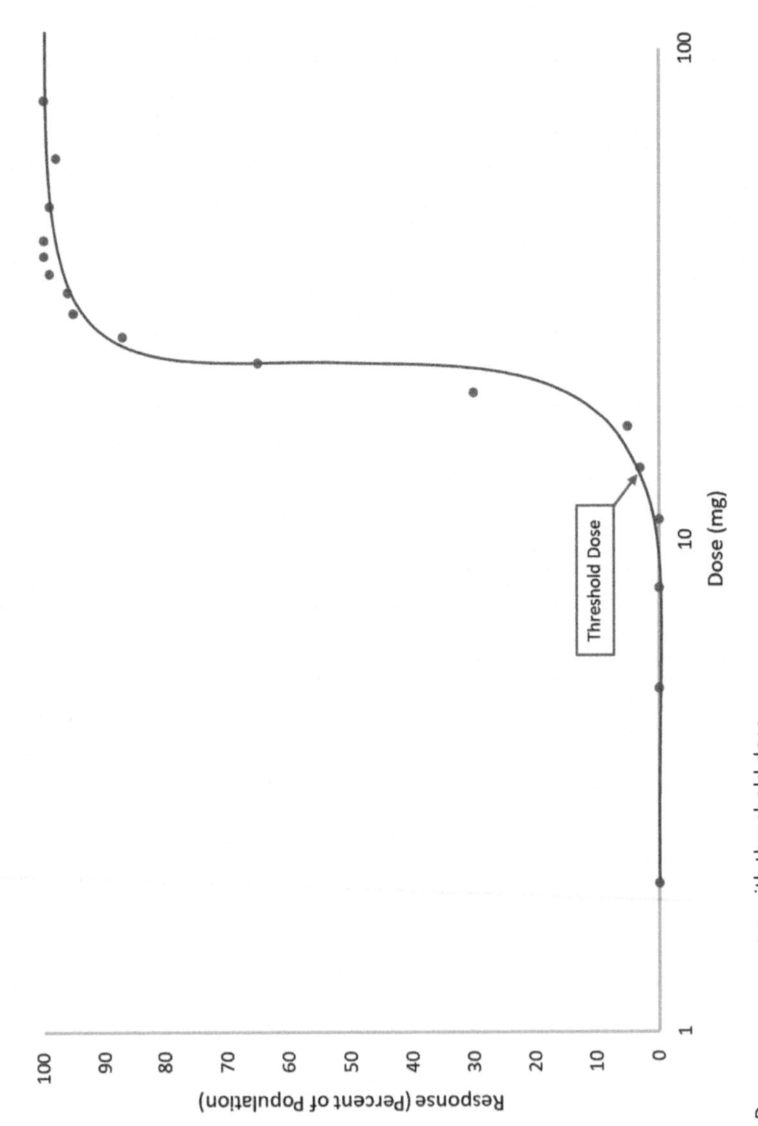


**Figure 4.3** Dose-response curve with threshold dose.

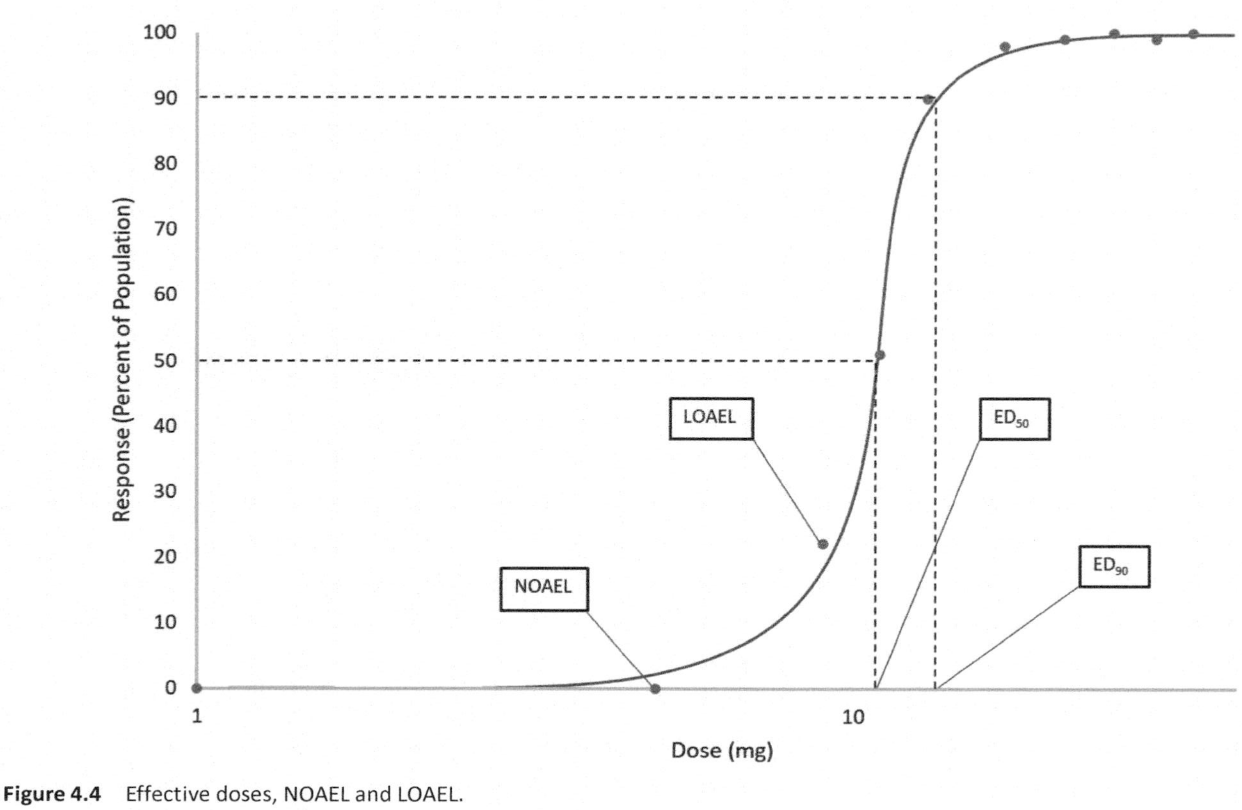


**Figure 4.4** Effective doses, NOAEL and LOAEL.

The effective dose (ED) is a factor used for estimating the response of the population to a substance exposure. Usually, it is expressed for 50% of the population as $ED_{50}$, but it can also be chosen at different levels, e.g., $ED_{10}$ or $ED_{90}$. In other words, $ED_{50}$ represents that at a specified dose, 50% of a population will exhibit an effect.

The toxic dose (TD) indicates the doses that will cause an adverse effect, and similar to the effective dose, it is usually expressed for 50% of the population ($TD_{50}$).

For pharmaceuticals, the relative safety of a drug is defined by the therapeutic index (TI), i.e., the rate of the TD to the therapeutically ED (Fig. 4.5). Commonly, the TI is derived from the $TD_{50}$ and $ED_{50}$ values. For example, clinicians will consider a drug is safer if it has a $TD_{50}/ED_{50}$ value of 10 compared to a value of 2. However, this value does not take into consideration the slope of the dose-response curve and the threshold values, e.g., for some substances a large increase in dose causes a small increase in response (shallow slope), while for others, a small increase in the dose will cause a large response (steep slope) (Fig. 4.6).

Another indicative way to estimate the adverse effects of chemicals often used in research and risk assessment is to establish the lowest dose at which a toxic or adverse effect is observed (lowest-observed-adverse-effect level, LOAEL), and the highest dose at which no toxic effects are identified (no-observed-adverse-effect level, NOAEL) (Fig. 4.4). For example, these indicators are used to estimate the maximum safe starting dose of a drug tested in clinical trials.

Several factors can affect the toxic and clinical effects of a dose. These are connected with the exact individual being exposed to the chemical, the time of exposure, and the concentration of the chemical. The individual's species, age, sex, and body size all influence the effects that a certain dose will have on that individual. For example, the same dose of a toxic chemical will not have the same effect on a shrew and an elephant. This is the reason why pharmaceuticals are administered in different dosages for children and adults. The time over which a dose is administered is especially important when the exposures are (sub)chronic. Commonly, chronic and subchronic doses are expressed using the time interval of 1 day, as mg/kg/day.

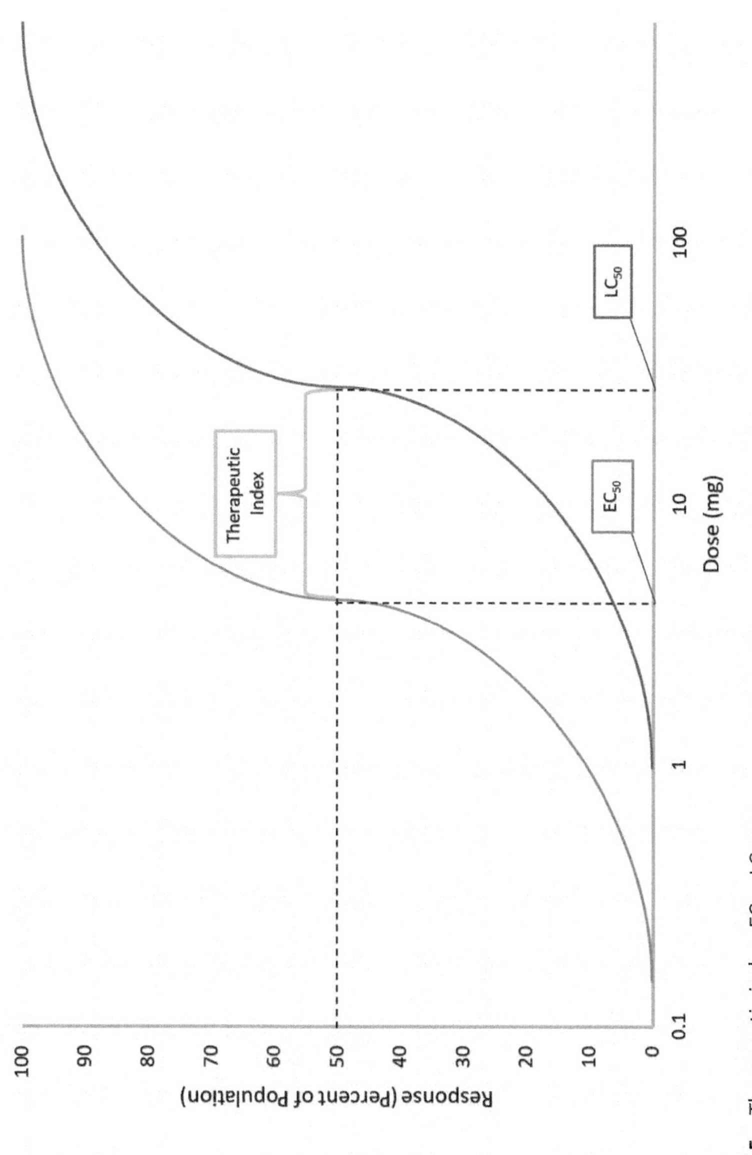


**Figure 4.5** Therapeutic index, $EC_{50}$, $LC_{50}$.

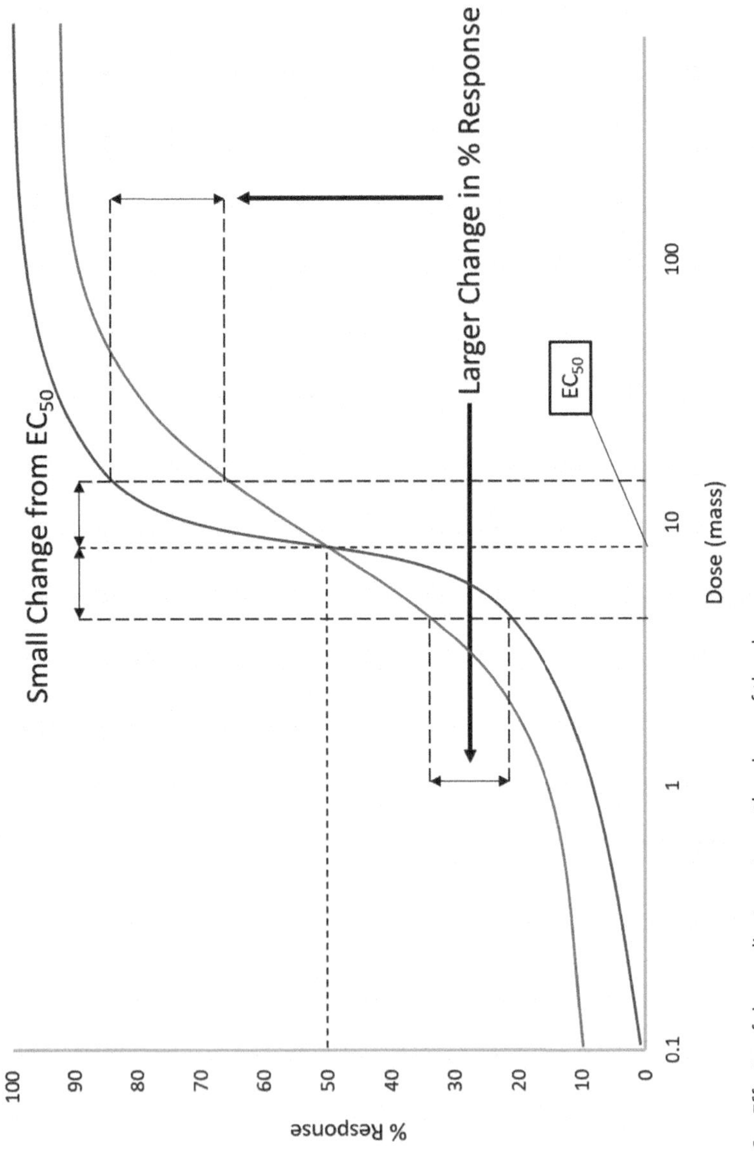


**Figure 4.6** Effects of dose adjustments on the slope of the dose-response curve.

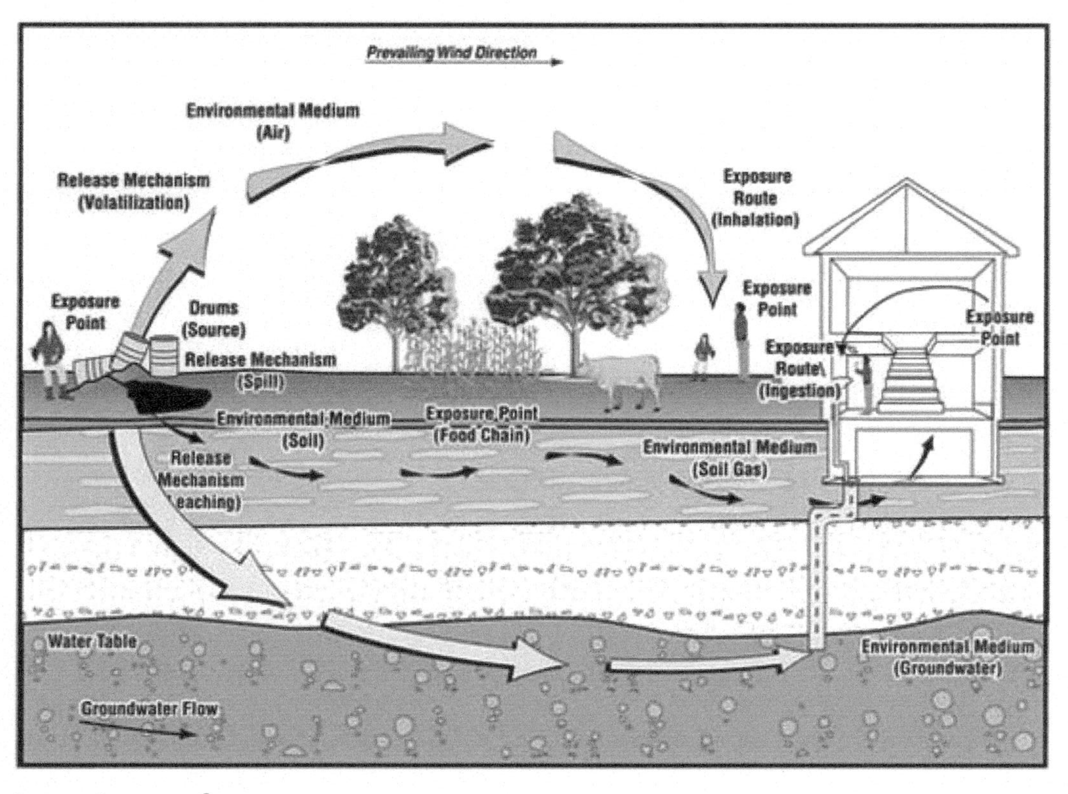


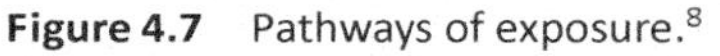
**Figure 4.7** Pathways of exposure.[8]

## 4.5 What Are the Pathways of Exposure?

As mentioned earlier in Chapter 2 (ADME, Section 2.1), once the individual contacts a chemical, it can get into the body following three common routes: through the GI tract, lungs, and skin. The exposure pathway, on the other hand, represents the physical route the chemical takes from the source to contact the individual. Commonly, exposure to a chemical originates from contact with contaminated air, food, soil, sediment, dust, and ground or surface water (Fig. 4.7).[7]

Sometimes, the contaminants can be found in multiple compartments, such as pesticides that can contaminate the air, crops, and soil.

To assess the exposure pathway, it is often necessary to consider several factors:[9]

(a) contaminant source, i.e., determine the place where a chemical is coming from
(b) environmental fate and transport of a chemical
(c) specific locations where the individual can come into contact with the chemical in the environment
(d) exposure route, i.e., identify the route a chemical takes to get into the body
(e) exposed population, i.e., determine the specific populations, and estimate the numbers that can be affected.

## References

1. Klaassen, C. D.; Watkins, III, J. B. *Casarett & Doull's Essentials of Toxicology*, 3rd ed., McGraw-Hill Education: New York, 2015, p. 524.
2. Sting. Mithridates IV Louvre.jpg. Wikimedia Commons, 2019.
3. Griffin, J. P. Famous names in toxicology. Mithridates VI of Pontus, the first experimental toxicologist. *Adverse Drug React Toxicol Rev*, 1995, **14**(1), 1–6.
4. Hinrichs, M. J. M.; Ryan, P. M.; Zheng, B.; Afif-Rider, S.; et al. Fractionated dosing improves preclinical therapeutic index of pyrrolobenzodiazepine-containing antibody drug conjugates. *Clin Cancer Res*, 2017, **23**(19), 5858–5868.

5. (a) Eaton, D. L.; Gilbert, S. G. Principles of toxicology. In: Klaassen, C. D.; Watkins, III, J. B. (Eds.), *Casarett & Doull's Essentials of Toxicology*, 3rd ed., McGraw-Hill Education: New York, 2015, pp. 5–20; (b) Calabrese, E. J. Dose–response relationship. In: Wexler, P. (Ed.), *Encyclopedia of Toxicology*, 3rd ed., Academic Press: Oxford, 2014, pp. 224–226.

6. (a) Balali-Mood, M.; Naseri, K.; Tahergorabi, Z.; Khazdair, M. R.; et al. Toxic mechanisms of five heavy metals: mercury, lead, chromium, cadmium, and arsenic. *Front Pharmacol*, 2021, **12**(227); (b) Clewell, R. A.; Thompson, C. M.; Clewell, H. J. Dose-dependence of chemical carcinogenicity: biological mechanisms for thresholds and implications for risk assessment. *Chem-Biol Interact*, 2019, **301**, 112–127.

7. Sunderland, E. M.; Hu, X. C.; Dassuncao, C.; Tokranov, A. K.; et al. A review of the pathways of human exposure to poly- and perfluoroalkyl substances (PFASs) and present understanding of health effects. *J Expo Sci Environ Epidemiol*, 2019, **29**(2), 131–147.

8. Agency for Toxic Substances and Disease Registry. *Public Health Assessment Guidance Manual* (Update), U.S. Department of Health and Human Services: Atlanta, Georgia, 2005, p. 357.

9. National Research Council (US) Commission on Engineering and Technical Systems; National Research Council (US) Commission on Life Sciences; McKone, T. E.; Huey, B. M.; et al. (Eds.), *Strategies to Protect the Health of Deployed U.S. Forces: Detecting, Characterizing, and Documenting Exposures*, The National Academies Press: Washington, DC, 2000.

Chapter 5

# Pharmacodynamics

## 5.1 What Is Pharmacodynamics?

Pharmacodynamics[1] is the study of biochemical, molecular, and physiologic actions or effects of a drug or other xenobiotic, in other words, the effect a chemical has on the body. The effect of a xenobiotic is expressed after the interaction with biological targets at the molecular level. These interactions, i.e., chemical interactions, receptor binding, and post-receptor effects induce a change in the target molecule, thus affecting the function and further intermolecular interactions.

A xenobiotic will exhibit the effect after the interaction with biologic targets, which can be direct or indirect and immediate or delayed. These effects differ in the mechanism of action and the pathway the chemical takes to exhibit its effect. Direct effects are those in which a xenobiotic directly interacts with the enzyme or receptor exhibiting the effect, while indirect effects are induced by chemicals that interact with biological targets resulting in a cascade of events leading to the final effect. Immediate effects are exhibited just after the interaction of the xenobiotic with the biological target, while delayed effects only emerge after some time passes after the initial interaction. Thus, pharmacodynamic studies are crucial in both toxicological and medicinal studies, establishing the harmful or beneficial doses/concentrations of a chemical.

*First Do No Harm: A Chemist's Guide to Molecular Design for Reduced Hazard*
Predrag V. Petrovic and Paul T. Anastas

ISBN 978-981-4968-59-1 (Hardcover), 978-1-003-35964-7 (eBook)
www.jennystanford.com

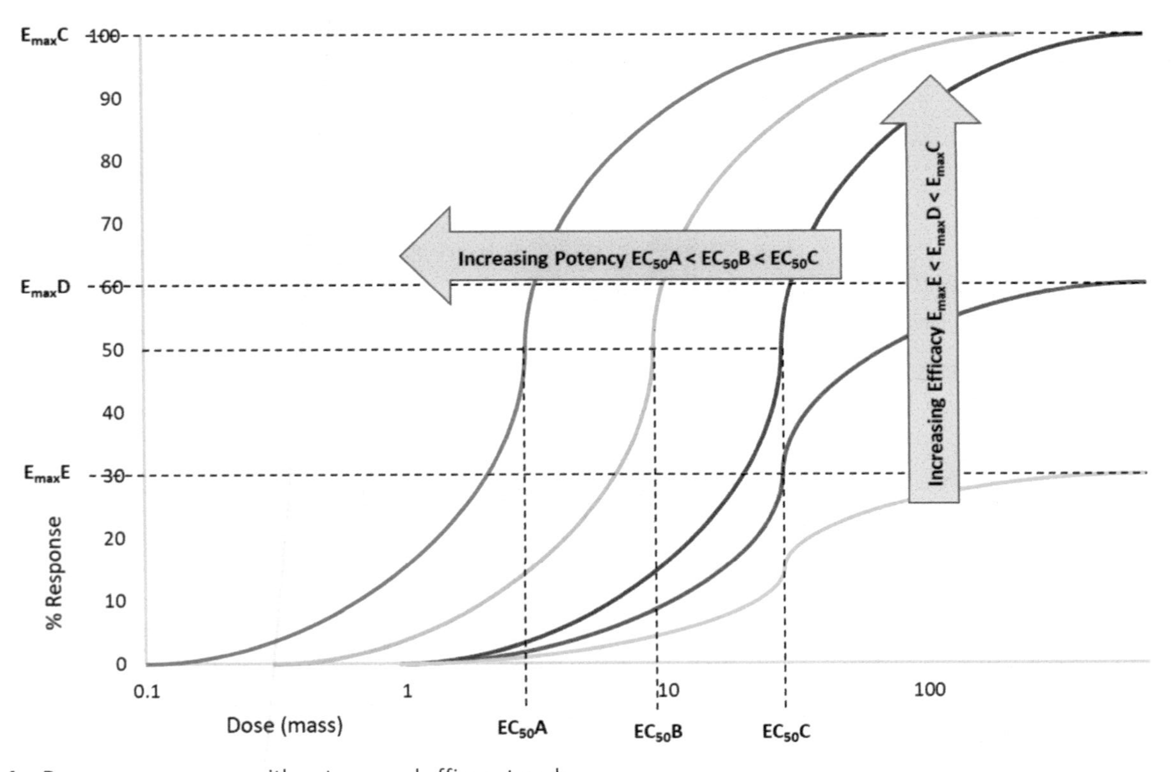


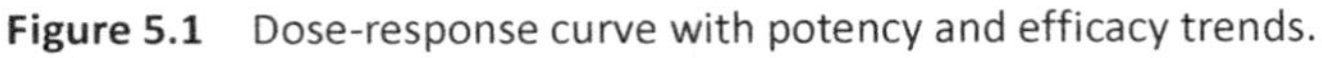

**Figure 5.1** Dose-response curve with potency and efficacy trends.

The effects of xenobiotics are usually measured by biochemical or clinical methods, and the extent and duration of the xenobiotic action are generally described by several parameters that can be derived from the dose-response curve: $EC_{50}$, $E_{max}$, and Hill coefficient (Fig. 5.1). $EC_{50}$ is the concentration of the xenobiotic that produces half of the maximum effect, $E_{max}$ is the maximal effect of the xenobiotic on the measured parameter, and Hill coefficient is the slope of the relationship between xenobiotic concentration and its effect.

## 5.2 What Are Receptor Interactions?

When the chemical enters the body and gets transported to the target cell, it can interact with the receptors, i.e., protein molecules located inside, or at the surface of the cell.[1a,2] These receptors have evolved to respond to a different group of molecules, such as hormones, neurotransmitters, and antigens, but various xenobiotic substances can also cause a response. When a chemical, i.e., a ligand, binds to a specific site at the receptor, it triggers a response in a cell. The primary receptors in the human body are proteins, however, xenobiotics can also bind to other biological targets: enzymes, ion channels, and transporters. While the changes initiated by the ligand-receptor complex can be direct or indirect, the ligand usually functions either as an agonist or antagonist (Fig. 5.2).[2]

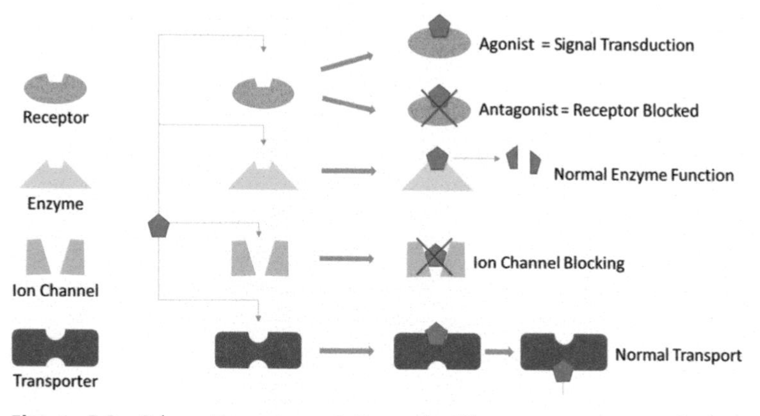


**Figure 5.2** Schematic representation of different receptors and their interactions.

When a chemical acts as an agonist, it imitates the molecule the receptor that it evolved to interact with, producing a similar response as the endogenous ligand. On the other hand, when acting as an antagonist, a chemical will block the interaction of the usual ligand with the receptor, inhibiting the related physiological response. The interactions of the ligand and receptor involve all types of bonding, i.e., covalent bonds, ionic bonds, hydrogen bonds, hydrophobic effect, and van der Waals interactions. For example, if the ligand binds to the receptor covalently, its action will last longer or potentially be irreversible, while non-covalent binding to the receptor usually results in a short duration of action. But mostly, the ligand and receptor interactions are the results of a combination of different bonding types. The strength and duration of interactions between the ligand and the receptor are mainly determined by the shape, structure, and reactivity of the ligand. The rate at which the association between ligand and receptor occurs compared to the rate of dissociation will provide insights into the affinity of the ligand to the receptor and $EC_{50}$ and $E_{max}$.

## 5.3 What Does the HOMO–LUMO Gap Has to Do with Pharmacodynamics?

In aquatic toxicity studies, frontier molecular orbital energies have been associated with the reactivity at the site of action.[3] These orbital energies, i.e., the highest occupied molecular orbital (HOMO) energy and lowest unoccupied molecular orbital (LUMO) energy, are obtained from quantum chemical computations and not experimentally derived. LUMO energy is usually a good predictor for electrophilic reactivity, but the difference between HOMO and LUMO energies, i.e., the energy gap $\Delta E$, was found to be a much better predictor of the acute aquatic toxicity than HOMO and LUMO energies by themselves.[4,5] Higher $\Delta E$ gaps were associated with decreased chemical reactivity but the study on the aquatic toxicity of the four species revealed that most chemicals with a $\Delta E$ value > 9 eV were generally "safe" and not toxic for the studied species. A smaller $\Delta E$ is generally associated with molecules that are considered chemically reactive for covalent bonding.[6] Frontier orbitals and the $\Delta E$ gap are also main parameters used in the conceptual density functional theory (DFT) to derive several descriptors that can indicate the reactivity of the molecule (Fig. 5.3).[7]

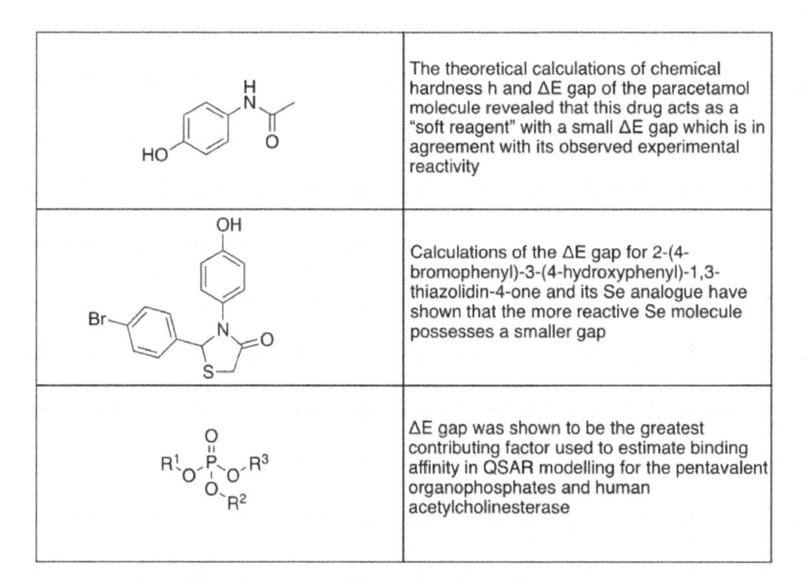

HO
H
N
O
The theoretical calculations of chemical hardness h and ΔE gap of the paracetamol molecule revealed that this drug acts as a "soft reagent" with a small ΔE gap which is in agreement with its observed experimental reactivity
OH
Br
N
O
S
Calculations of the ΔE gap for 2-(4-bromophenyl)-3-(4-hydroxyphenyl)-1,3-thiazolidin-4-one and its Se analogue have shown that the more reactive Se molecule possesses a smaller gap
O
R1 O P O R3
O
R2
ΔE gap was shown to be the greatest contributing factor used to estimate binding affinity in QSAR modelling for the pentavalent organophosphates and human acetylcholinesterase


**Figure 5.3** Frontier orbital energies application in describing molecular reactivity of selected chemical compounds.[8]

## References

1. (a) Currie, G. M. Pharmacology, Part 1: Introduction to pharmacology and pharmacodynamics. *J Nucl Med Technol*, 2018, **46**(2), 81–86; (b) Marino, M.; Jamal, Z.; Zito, P. M. *Pharmacodynamics*, StatPearls Publishing: Treasure Island, FL, 2020.

2. Rang, H. P.; Dale, M. M. *Rang & Dale's Pharmacology*, 6 ed., Churchill Livingstone: Edinburgh, 2007.

3. (a) Lewis, D. F. V. Frontier orbitals in chemical and biological activity: quantitative relationships and mechanistic implications*. *Drug Metab Rev*, 1999, **31**(3), 755–816; (b) Zvinavashe, E.; Murk, A. J.; Vervoort, J.; Soffers, A. E.; et al. Quantum chemistry based quantitative structure-activity relationships for modeling the (sub)acute toxicity of substituted mononitrobenzenes in aquatic systems. *Environ Toxicol Chem*, 2006, **25**(9), 2313–2321; (c) Voutchkova-Kostal, A. M.; Kostal, J.; Connors, K. A.; Brooks, B. W.; et al. Towards rational molecular design

for reduced chronic aquatic toxicity. *Green Chem*, 2012, **14**(4), 1001–1008.

4. Voutchkova, A. M.; Kostal, J.; Steinfeld, J. B.; Emerson, J. W.; et al. Towards rational molecular design: derivation of property guidelines for reduced acute aquatic toxicity. *Green Chem*, 2011, **13**(9), 2373–2379.
5. Corrales, J.; Kristofco, L. A.; Steele, W. B.; Saari, G. N.; Kostal, J.; Williams, E. S.; Mills, M.; Gallagher, E. P.; Kavanagh, T. J.; Simcox, N.; Shen, L. Q.; Melnikov, F.; Zimmerman, J. B.; Voutchkova-Kostal, A. M.; Anastas, P. T.; Brooks, B. W. Toward the design of less hazardous chemicals: Exploring comparative oxidative stress in two common animal models. *Chem Res Toxicol*, 2017, **30**(4), 893–904. http://dx.doi.org/10.1021/acs.chemrestox.6b00246
6. Kostal, J.; Voutchkova-Kostal, A.; Anastas, P. T.; Zimmerman, J. B. Identifying and designing chemicals with minimal acute aquatic toxicity. *Proc Natl Acad Sci*, 2015, **112**(20), 6289–6294.
7. Domingo, L. R.; Ríos-Gutiérrez, M.; Pérez, P. Applications of the conceptual density functional theory indices to organic chemistry reactivity. *Molecules*, 2016, **21**(6), 22.
8. (a) Ruark, C. D.; Hack, C. E.; Robinson, P. J.; Anderson, P. E.; et al. Quantitative structure-activity relationships for organophosphates binding to acetylcholinesterase. *Arch Toxicol*, 2013, **87**(2), 281–289; (b) Syrovaya, A. O.; Levashova, O. L.; Andreeva, S. V. Investigation of quantum-chemical properties of paracetamol. *J Chem Pharm Res*, 2015, **7**(1), 307–311; (c) Kavitha, H. P.; Rhyman, L.; Ramasami, P. Molecular structure and vibrational spectra of 2-(4-bromophenyl)-3-(4-hydroxyphenyl) 1,3-thiazolidin-4-one and its selenium analogue: insights using HF and DFT methods. *Phys Sci Rev*, 2019, **4**(3), 20180031.

# Chapter 6

# Classes of Chemicals

As we discussed through the earlier chapters, the design of chemicals was historically limited and focused on the function rather than on the whole lifecycle of the designed molecule. Understandably, this led to the numerous problems caused by inadvertent hazards these chemicals caused and that we must deal with nowadays. This chapter will cover several classes of chemicals that illustrate this point very well, ranging from organohalogens, through polymers, epoxides, nitriles, a majority of electrophiles, to asbestos.

## 6.1 What Are the Biggest Concerns About Organohalogens?

Organohalogens are a large class of organic compounds that contain one or more halogen atoms (fluorine, chlorine, and bromine). These chemicals possess physicochemical properties that make them extremely persistent in the environment, often bioaccumulating across many species which finally causes long-term effects to humans as well. Due to their persistence, a large number of synthetic organohalogens, such as polychlorinated biphenyls (PCBs), agricultural pesticides, or brominated flame retardants, have become global environmental contaminants, i.e., persistent organic pollutants (POPs) (Fig. 6.1).[1]

*First Do No Harm: A Chemist's Guide to Molecular Design for Reduced Hazard*
Predrag V. Petrovic and Paul T. Anastas

ISBN 978-981-4968-59-1 (Hardcover), 978-1-003-35964-7 (eBook)
www.jennystanford.com

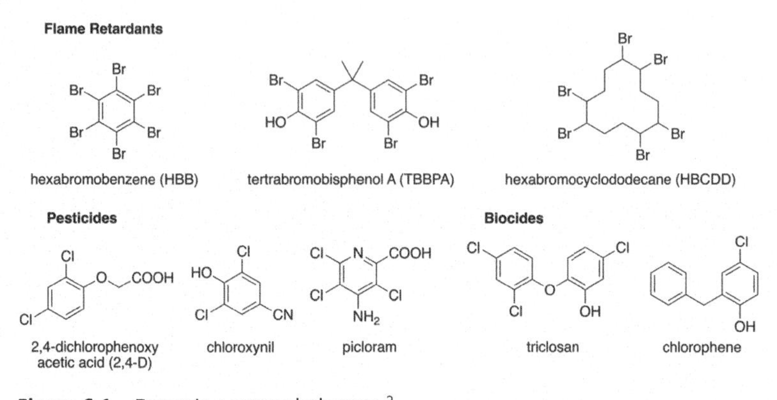

**Figure 6.1** Emerging organohalogens.[2]

Initially, PCB mixtures were introduced for a variety of applications such as additives in plastics, adhesives, paints, etc. and as hydraulic, lubricating, and heat-transfer fluids. Additionally, PCBs DDT, chlordane, and HCH have been used as very efficient insecticides beginning in the 1940s reducing the number of insect-borne diseases such as malaria, and their widespread application increased agricultural productivity all over the world. However, the decades of use revealed a number of environmental and human health issues caused by these PCBs' persistence, bioaccumulation, and toxicity, which led to the severe regulatory restrictions and bans of their use in developed countries in the 1970s. This change was sparked by Rachel Carson's *Silent Spring*,[3] which showed the effects of synthetic pesticides on the environment. The properties of a PCB depend on the number of chlorine atoms and their position in the molecule, but generally, these molecules have high thermal stability while being resistant to the influence of acids and bases. PCBs are also poorly soluble in water and readily soluble in biological lipids. All of these properties make them very useful for numerous industrial applications, but also causes persistence in the environment and various adverse health effects including endocrine system disruption, neurotoxicity, cancer, reproductive toxicity, etc.

Another group of organohalogen compounds that found wide application in a range of industrial and consumer products (starting in the 1950s with the introduction of teflon) are polyfluorinated compounds (PFCs).[4] These compounds have been used in surfactants, cleaning products, pharmaceuticals, herbicides, oil and water-repelling coatings, lubricants, polishers, emulsifiers, etc. PCFs are mostly amphiphilic and possess thermodynamically very strong C-F bonds that are responsible for their inherent chemical stability and resistance to degradation. However, this stability causes their extreme persistence and bioaccumulation in the environment, which has become evident in recent decades. The samples from the environment and even the human blood from around the world have shown the presence of PFCs such as perfluorooctane sulfonate (PFOS), perfluorooctanoic acid (PFOA) and others. The highest concentrations were usually found close to the areas where direct industrial emissions occurred. The PFCs have been connected with various health effects in animals such as liver enlargement, tumor cases increase, and change in metabolism, while studies on the effects on human health suggest changes in gene production and the immune system.

Finally, since the 1970s brominated compounds have seen increased use in a broad range of consumer products as flame retardants. The polybrominated biphenyls (PBBs), and polybrominated diphenyl ethers (PBDEs) are regularly used as additives in fabrics, foams, and plastics and can be found in a number of electronic devices and various furniture. The PBDEs possess relatively weak C-Br bonds, which are thermally labile resulting in the formation of $Br^{\bullet}$ radicals. These radicals are responsible for the decrease of the flame intensity and CO production by reacting with carbon radicals created in the burning process. Because of their prevalence in household items, these compounds are primarily indoor pollutants and are often found in house dust. However, they have also been found in the air, water, soil, sediments, aquatic organisms, other wildlife, and humans.[5] PBDEs are structurally very similar to PCBs and are also very resistant to both abiotic and biotic degradation, however, they are slightly more reactive and more prone to bioactivation than PCBs. Less brominated PBDEs such as tetra- and penta-PBDE have been found in breast milk and other fat tissues, as they have a high affinity for lipids. They have also been associated with thyroid hormone imbalances, neurodevelopmental toxicity, and tumors, and are suspected endocrine disruptors.[6]

## 6.2 What Are the Biggest Concerns for Epoxides?

Epoxides are a class of chemicals that contain a cyclic ether moiety consisting of a three-atom ring. This system is highly strained, and thus very reactive. This reactivity and their electrophilic and lipophilic nature are the main characteristics that can potentially cause adverse effects in the body. Ring-opening reactions are initiated by nucleophiles containing $OH^-$, $S^-$, $Cl^-$, or $NH_2$ groups, forming the products as shown in Fig. 6.2.

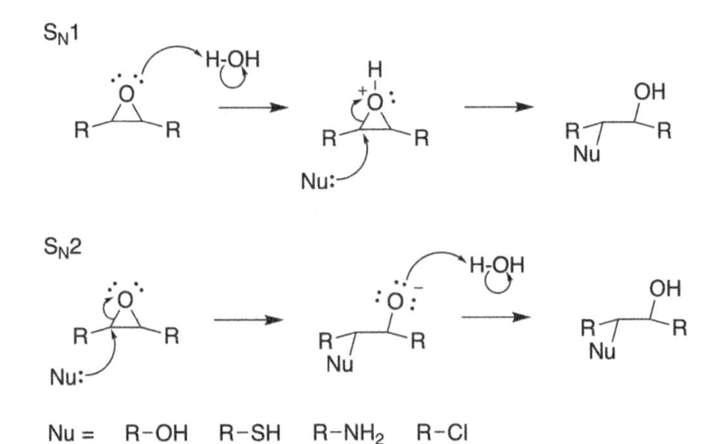

**Figure 6.2** Epoxide reactivity and typical chemical reactions in the metabolism.

Epoxides are a large class of chemicals that are produced in large quantities by industry, but they also occur naturally in the environment (e.g., aflatoxin) and are generated as metabolites in some biochemical processes.[7] They occur as gases, liquids, and solids and are very useful as crosslinking and alkylating agents due to their reactivity. While some epoxides are used by themselves (e.g., sterilizing agents), most are used as intermediates in various products such as adhesives, specialized chemicals, cement, surface-active agents, or synthetic resins. Often, they are also used as stabilizers and plasticizers in polymers.

Due to their high reactivity and electrophilicity, epoxides are considered potential carcinogens and mutagens.[9] They are usually detoxified in the metabolism via enzymatic and non-enzymatic reactions, but if not removed, they can covalently bind with nucleophilic centers on proteins and nucleic acids (Fig. 6.3). This binding can in turn cause toxic effects such as carcinogenesis or mutagenesis. For example, the phase II enzyme glutathione (GSH) transferase protects cellular targets from electrophiles through its nucleophilic SH group. Epoxides are conjugated with glutathione via one of the carbons in the ring by opening the ring. This stable conjugate is further metabolized and subsequently excreted from the body via urine.

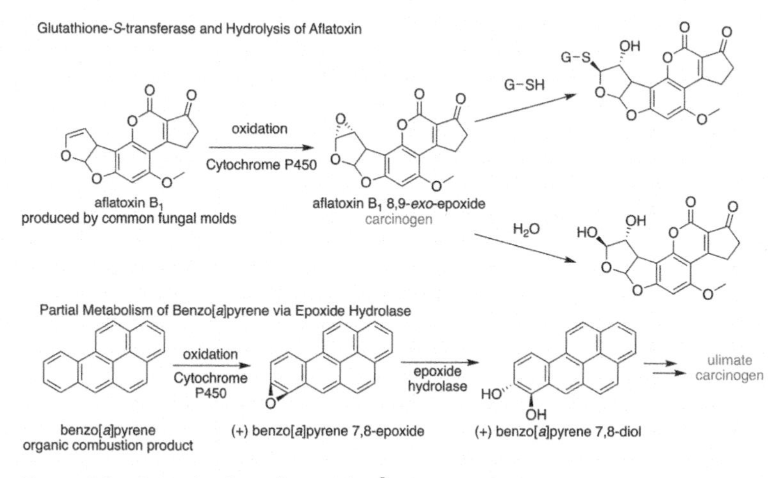


**Figure 6.3** Detoxication of epoxides.[8]

## 6.3 What Are the Biggest Concerns for Polymers?

Synthetic polymers have become one of the staples of modern society, a trend initiated by Bakelite—the first mass-produced synthetic polymer at the beginning of the 20th century.[10] Later, other synthetic polymers (e.g., polyvinylchloride, polyurethane, polystyrene, polypropylene, etc.) were developed, leading to the rapid expansion of the polymer industry. Nowadays, most of these polymers have become a significant environmental issue due to their inherent resistance to degradation and persistence in the environment, along with the release of numerous hazardous substances when degradation occurs, due to weathering for example (Table 6.1).[11] These substances, such as plasticizers, stabilizers, flame retardants, and lubricants are often added to the polymers to achieve desired physical and chemical properties, improve processability, and prolong life span.

**Table 6.1** Most commonly used synthetic polymers and their hazard potential

| Name | Description |
|---|---|
| Polyvinylchloride (PVC) | • possesses a high concentration of chlorine atoms<br>• easily degrades by the influence of light or heat, and reaction is additionally catalyzed by released HCl<br>• phthalates are added as plasticizers, while toxic transition metal additives are used to improve stability |
| Polycarbonates | • produced with bisphenol A as part of the chain or additive which is released when decomposed<br>• biodegradation is slow because of the presence of aromatic rings |
| Polyolefins | • contain short and long branches making it accessible for attacks by free radicals and oxidase enzymes; branching slows the biotic degradation<br>• produced from non-renewable sources by polymerizing alkenes<br>• used for a large number of flexible consumer packaging products making them widespread<br>• can contain toxic pigments, antiacid and antioxidant additives, and catalytic residues |
| Polyethylenes (PET, LDPE, HDPE) | • LDPE and HDPE have a density higher than water, so they float on the surface until decomposed<br>• hydrolysis of ester group in PET accelerated by UV radiation, acids, bases, and some transition metals<br>• often polluted by the presence of metallic catalysts and phosphorus compounds |
| Polystyrene | • presence of aromatic rings reduces biodegradation<br>• often smaller fragments (dimers, trimers) are volatile; phenyl group absorbs UV radiation creating radicals in the process |
| Polyurethane (PUR) | • polyethers and polyesters that contain isocyanate groups and release volatile organic contaminants |

Polymers are chemical compounds consisting of a large number of repeating subunits, i.e., monomers, that build a large macromolecular structure. However, polymers are rarely pure compounds, and often contain different additives (e.g., phthalate or antioxidant plasticizers) introduced to either protect the polymer from different types of degradation or to improve their properties; and various impurities as residues from the synthesis process (e.g., catalysts or solvents). The typical molecular weight of polymers ranges from 30,000 Da to more than 1,000,000 Da, and most polymers are hydrophobic. Both of these properties, along with the presence of aromatic rings and chlorine atoms in some of the most common polymers, contribute to their limited biodegradation.

Generally, synthetic polymers can be classified as either thermosets or thermoplastics (Fig. 6.4).

(a) Thermosets create 3D networks between the polymer chains when pressure and heat are applied, thus hardening and taking permanent shape. Typical examples of this group are epoxy resins, most polyurethanes, and Bakelite.

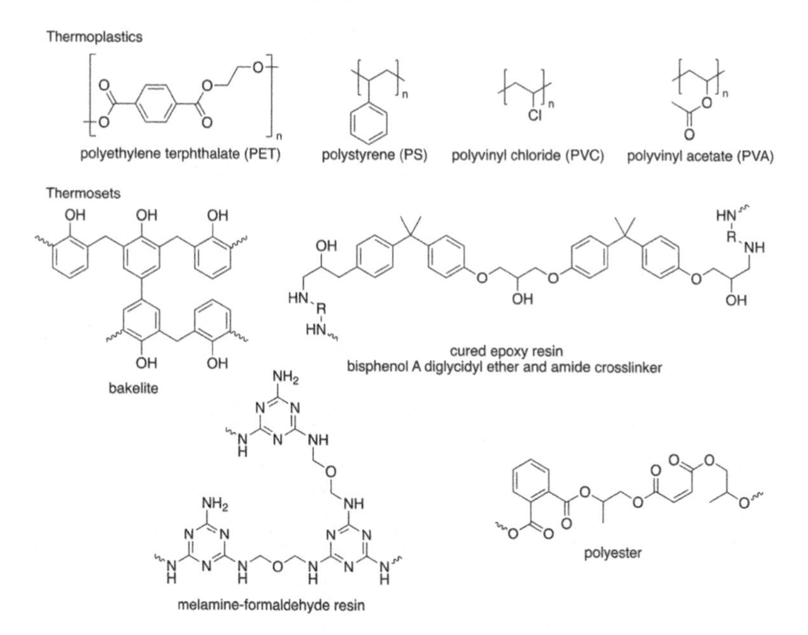


**Figure 6.4** Chemical structures of some thermoplastic and thermoset polymers.

(b) Thermoplastics are the most common group of synthetic polymers. As their name implies, these polymers can be repeatedly melted by heating and solidified by cooling, taking different desired shapes.

Polymers are mainly degraded via three routes, oxidatively, hydrolytically, and enzymatically.

**(a) Oxidative degradation** is initiated by the formation of free radicals in the polymer chain in the presence of oxygen. These radicals can be formed under the influence of heat, UV radiation, or mechanical deformation. This process occurs naturally, with the breakage of the molecular chains into smaller fragments where the oxygen is incorporated as different functional groups (e.g., carbonyl, ether, hydroxyl, etc.). However, the complete degradation of a polymer can take a very long time, i.e., decades, or centuries, unless some additives, such as transition metals, are present.

**(b) Hydrolytic degradation** of polymers involves the attack of water molecules that may break amide, ether, or ester bonds in a chain. This reaction is often catalyzed by acids and bases and results in the formation of amines and carboxylic acids (amides), two alcohols (ethers), or alcohols and carboxylic acids (esters). The hydrolysis of esters is a much easier process compared to the hydrolysis of ethers, which often causes greater persistence of the polymers containing ether bonds.

**(c) Enzymatic degradation**, i.e., biodegradation, is a process where polymer chains are broken by the action of various enzymes, e.g., esterases degrade polyesters. These enzymes are responsible for the natural degradation of many biopolymers such as cellulose and starch but can be involved in the degradation of synthetic polymers as well.

## 6.4 What Are the Biggest Concerns with Nitriles?

Nitriles groups can be found in a large number of xenobiotics and pharmaceutical drugs, and additionally, aliphatic and aromatic nitriles are widely used in the manufacturing of various solvents and plastics.[12] The presence of nitrile groups in molecules make them potential hazards as they can be metabolized by the cytochrome P450 enzyme which results in the release of very toxic cyanide (Fig. 6.5). Some nitriles are known to cause adverse effects on the health of various organisms.

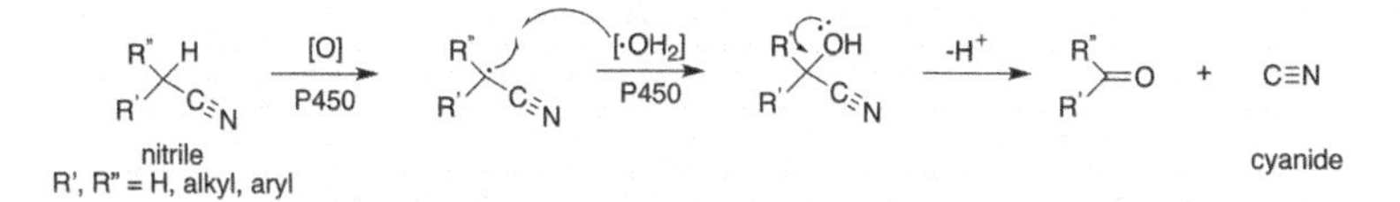


**Figure 6.5** Mechanism of the CYP450-mediated release of cyanide from nitriles.[13]

CYP450 catalyzes the hydroxylation of nitriles, which leads to oxygen activation, hydrogen atom abstraction, recombination of resulting hydroxy radical with the substrate, and finally dissociation of hydroxylated product, i.e., cyanohydrin, which quickly releases hydrogen cyanide.[13]

Since the oxidation of nitriles by CYP450 is based on mechanisms involving radical species, nitrile toxicity can be approximated by observing the structural features influencing the stability of the nitrile radicals. The rate by which the hydrogen atom is abstracted from α-carbon is related to the rate of cyanohydrin (and cyanide) formation. For example, a nitrile group bound to the primary carbon atom forms less stable radicals than one bound to secondary, tertiary, or benzylic carbon; thus, acetonitrile is considered less toxic than propionitrile or isobutyronitrile.[14] Still, all aliphatic nitriles are considered toxic since released cyanide causes degeneration of neurons and blocks ATP production which can ultimately cause acute lethality.[15]

## 6.5 What Are the Biggest Concerns with Electrophiles?

Electrophiles are a class of chemicals that are electron-deficient and thus susceptible to interactions with nucleophilic centers resulting in the formation of covalent bonds. While reversible interactions of electrophiles and nucleophiles are normal processes in the cell (e.g., enzyme regulation, or release of neurotransmitters), electrophiles can irreversibly bind to macromolecules which can cause various adverse effects.[16] However, the exact mechanism of electrophile toxicity is still not entirely known since it cannot be defined by only one mechanism of action.[17] The main mechanisms of electrophile interactions with biological nucleophiles are additions to carbon-oxygen or carbon-carbon double bonds and nucleophilic displacement reactions. Electrophiles can react with different types of nucleophilic centers but most relevant for the biological systems are amino ($NH_2$), hydroxy (OH), and thiol (SH) functional groups, since they are common constituents of proteins and organic bases in DNA.

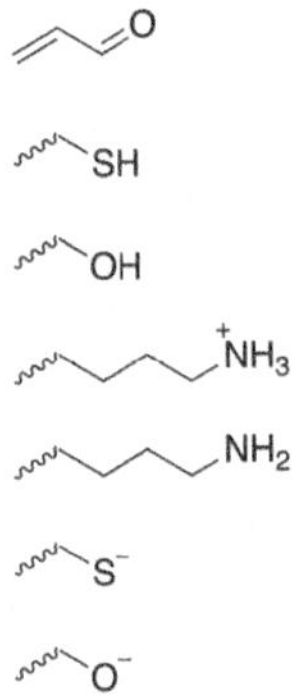

| | | Softness ($\sigma$, $eV^{-1}$) | Hardness ($\eta$, eV) | Nucleophilicity with Acrolein ($\omega$–, eV) |
|---|---|---|---|---|
| acrolein (food contaminant) | O | **0.372** | **2.69** | — |
| cysteine | SH | 1.724 | 0.58 | 0.050 |
| serine | OH | 0.289 | 3.46 | 0.061 |
| lysine cation | $\overset{+}{N}H_3$ | 0.259 | 3.86 | 0.063 |
| lysine | $NH_2$ | 0.296 | 3.38 | 0.083 |
| cysteine anion | $S^-$ | 0.601 | 1.67 | 0.639 |
| serine anion | $O^-$ | 0.373 | 2.68 | 0.979 |

**Figure 6.6** Reactions of electrophiles with nucleophiles.[16]

One way of classifying the components of electrophiles and nucleophiles contributing to their covalent interactions is by using Pearson's Hard and Soft Acids and Bases (HSAB) theory.[18] This concept is based on the calculation of quantum chemical parameters from the electronic properties that describe their reactivity. HSAB theory uses the delocalization of electron density to form covalent bonds, i.e., polarizability, to classify electrophiles and nucleophiles as either "soft" or "hard". This implies that covalent adducts will be formed by electrophiles and nucleophiles of comparable softness or hardness. Elements with smaller atomic radii have outer-shell electrons closer to the nucleus compared to the elements with large radii, making their electron clouds harder to distort. Thus, the smaller $NH_2$, and OH groups will be less polarizable, and thus "harder" than the SH group which will act as a "soft" nucleophile. According to the HSAB principles, hard electrophiles will interact with hard nucleophiles, while soft electrophiles will interact with soft nucleophiles. Some examples of reactive electrophiles reactions with the SH group are shown in Fig. 6.6.

## 6.6 What Are the Biggest Concerns with Asbestos?

Asbestos is a naturally occurring material made of fibrous silicate minerals bonded with aluminum, calcium, iron, magnesium, and sodium, and it possesses resistance to heat, electricity, and corrosion. While these properties made asbestos fibers very useful for various commercial applications, they also make asbestos very stable and persistent in the environment. Exposure to the fibers can cause severe toxicity.[19] Asbestos is classified by the type of fibers as either amphibole, with brittle needle-shaped fibers; and serpentine, with long and flexible fibers (Table 6.2).

Asbestos (Fig. 6.7) was used in commercial and industrial applications widely because it is a very flexible material, possesses high tensile strength, and is resistant to corrosion (by acids and bases) and heat and flame. Thus, it was often used as an insulating material by itself, or as an additive to improve the strength of cement, clothing, paper, plastic, and other materials.

However, over decades of use, it became evident that any exposure to it causes various health issues, such as different types of cancers (lung, laryngeal, ovarian). Mesothelioma is almost exclusively caused by asbestos, but it can also cause other types of diseases such as asbestosis, pleuritis, pleural effusions, etc. Some of the adverse effects come from the inhalation of the asbestos fibers, which can penetrate deep into the lungs and get lodged in the tissue, causing alveolar and interstitial fibrosis.

**Table 6.2** Common asbestos minerals, and asbestos fiber[20]

| Common Name | Mineral Group | Idealized Chemical Formula | Color | Decomp. Temp. (°C) |
|---|---|---|---|---|
| **Chrysotile** | Serpentine | $[Mg_3Si_2O_5(OH)_4]_n$ | White, grey, green, yellowish | 600–850 |
| **Crocidolite** | Amphibole | $[NaFe^{2+}{}_3Fe^{3+}{}_2Si_8O_{22}(OH)_2]_n$ | Lavender, blue-green | 400–900 |
| **Amosite** | Amphibole | $[(Mg, Fe^{2+})_7Si_8O_{22}(OH)_2]_n$ | Brown, grey, greenish | 600–900 |
| **Anthophyllite** | Amphibole | $[(Mg, Fe^{2+})_7Si_8O_{22}(OH)_2]_n$ | Grey, white, brown-grey, green | Not Reported |
| **Actinolite** | Amphibole | $[Ca_2(Mg, Fe^{2+})_5Si_8O_{22}(OH)_2]_n$ | Green | Not Reported |
| **Tremolite** | Amphibole | $[Ca_2Mg_5Si_8O_{22}(OH)_2]_n$ | White to pale green | 950–1040 |

Figure 6.7 Anthophyllite asbestos.[21]

## References


1. Kodavanti, P. R. S.; Loganathan, B. G. Organohalogen pollutants and human health. In: Quah, S. R. (Ed.), *International Encyclopedia of Public Health*, 2nd ed., Academic Press: Oxford, 2017, pp. 359–366.
2. Calvo-Flores, F. G.; Isac-Garcia, J.; Dobado, J. A. *Emerging Pollutants: Origin, Structure, and Properties*. John Wiley & Sons: Newark, 2018.
3. Carson, R. *Silent Spring*. Houghton Mifflin: Boston, Cambridge, MA, 1962, p. 368.
4. Lindstrom, A. B.; Strynar, M. J.; Libelo, E. L. Polyfluorinated compounds: past, present, and future. *Environ Sci Technol*, 2011, **45**(19), 7954–7961.
5. Kerrigan, J. F.; Engstrom, D. R.; Yee, D.; Sueper, C.; et al. Quantification of hydroxylated polybrominated diphenyl ethers (OH-BDEs), triclosan, and related compounds in freshwater and coastal systems. *PLoS One*, 2015, **10**(10), e0138805.

6. Siddiqi, M. A.; Laessig, R. H.; Reed, K. D. Polybrominated diphenyl ethers (PBDEs): new pollutants-old diseases. *Clin Med Res*, 2003, **1**(4), 281–290.
7. Manson, M. M. Epoxides: is there a human health problem? *Br J Ind Med*, 1980, **37**(4), 317–336.
8. (a) Johnson, W. W.; Ueng, Y.-F.; Widersten, M.; Mannervik, B.; et al. Conjugation of highly reactive aflatoxin B1 exo-8,9-epoxide catalyzed by rat and human glutathione transferases: estimation of kinetic parameters. *Biochemistry*, 1997, **36**(11), 3056–3060; (b) Shimada, T. Xenobiotic-metabolizing enzymes involved in activation and detoxification of carcinogenic polycyclic aromatic hydrocarbons. *Drug Metab Pharmacokinet*, 2006, **21**(4), 257–276.
9. Kostal, J.; Voutchkova-Kostal, A.; Weeks, B.; Zimmerman, J. B.; et al. A free energy approach to the prediction of olefin and epoxide mutagenicity and carcinogenicity. *Chem Res Toxicol*, 2012, **25**(12), 2780–2787.
10. Vlachopoulos, J.; Strutt, D. Polymer processing. *Mater Sci Technol*, 2003, **19**(9), 1161–1169.
11. (a) Ojeda, T. Polymers and the environment. In: Yılmaz, F. (Ed.), *Polymer Science*, IntechOpen, 2013; (b) Ahmad, A. F.; Razali, A. R.; Razelan, I. S. M. Utilization of polyethylene terephthalate (PET) in asphalt pavement: a review. *IOP Conf Ser: Mater Sci Eng*, 2017, **203**, 012004.
12. Tanii, H.; Hashimoto, K. Studies on the mechanism of acute toxicity of nitriles in mice. *Arch Toxicol*, 1984, **55**(1), 47–54.
13. Grogan, J.; DeVito, S. C.; Pearlman, R. S.; Korzekwa, K. R. Modeling cyanide release from nitriles: prediction of cytochrome P450 mediated acute nitrile toxicity. *Chem Res Toxicol*, 1992, **5**(4), 548–552.
14. DeVito, S. C. Structural and toxic mechanism-based approaches to designing safer chemicals. In: Anastas, P. T.; Boethling, R.; Voutchkova, A. (Eds.), *Handbook of Green Chemistry*, Wiley-VCH: Weinheim, Germany, 2012, pp. 77–106.
15. Moser, V. C.; Aschner, M.; Richardson, R. J.; Philbert, M. A. Toxic responses of the nervous system. In: Klaassen, C. D.; Watkins, III, J. B. (Eds.), *Casarett & Doull's Essentials of Toxicology*, 3rd ed., McGraw-Hill Education: New York, 2015, pp. 237–254.
16. (a) LoPachin, R. M.; Gavin, T. Reactions of electrophiles with nucleophilic thiolate sites: relevance to pathophysiological mechanisms and remediation. *Free Radic Res*, 2016, **50**(2), 195–205; (b) LoPachin, R. M.; Geohagen, B. C.; Nordstroem, L. U. Mechanisms of soft and hard electrophile toxicities. *Toxicology*, 2019, **418**, 62–69.

17. Hermens, J. L. Electrophiles and acute toxicity to fish. *Environ Health Perspect*, 1990, **87**, 219–225.

18. Pearson, R. G. Hard and soft acids and bases: the evolution of a chemical concept. *Coord Chem Rev*, 1990, **100**, 403–425.

19. (a) Agency for Toxic Substances and Disease Registry. *Case Studies in Environmental Medicine: Asbestos Toxicity*, U.S. Department of Health and Human Services, 2014; (b) Rosenthal, G. J.; Simeonova, P.; Corsini, E. Asbestos toxicity: an immunologic perspective. *Rev Environ Health*, 1999, **14**(1), 11–20.

20. International Agency for Research on Cancer. *Arsenic, Metals, Fibres, and Dusts: A Review of Human Carcinogens*, International Agency for Research on Cancer: Lyon, France, 2012, p. 527.

21. Survey, U. S. G. anthophyllite asbestos, Georgia (public domain). https://www.usgs.gov/media/images/anthophyllite-asbestos-0 (accessed December 15).

Chapter 7

# Design Rules for Safer Chemicals

## 7.1 Strategies to Minimize Hazard

Designing safer chemicals is a complicated process that requires careful consideration due to the vastness of different factors that are all intrinsically interconnected. Some general approaches can still be used as a sound strategy when considering the initial stages of the design process. To minimize the hazard potential of chemicals these key steps should be attempted:

(a) Bioavailability should be either eliminated or at least reduced
(b) The use of dangerous additives should be reduced
(c) Reactive functional groups should be replaced or modified
(d) Chemicals should be designed for degradation after their functional life ends
(e) Biological pathways of action should be terminated or at least modified

When thinking about these strategies used in molecular design, a helpful concept that can be effectively applied is the toxicological hazard face of the molecular design pyramid shown in Fig. 7.1. The pyramid is constructed from blocks representing different tools that can be used in molecular design and are arranged in increased complexity starting from the simplest at the base, with the more

*First Do No Harm: A Chemist's Guide to Molecular Design for Reduced Hazard*
Predrag V. Petrovic and Paul T. Anastas

ISBN 978-981-4968-59-1 (Hardcover), 978-1-003-35964-7 (eBook)
www.jennystanford.com

intricate concepts towards the top. While there are a lot of complex issues that cannot be easily manipulated due to our current lack of understanding, today we can still use numerous simple tools towards safer chemical design.

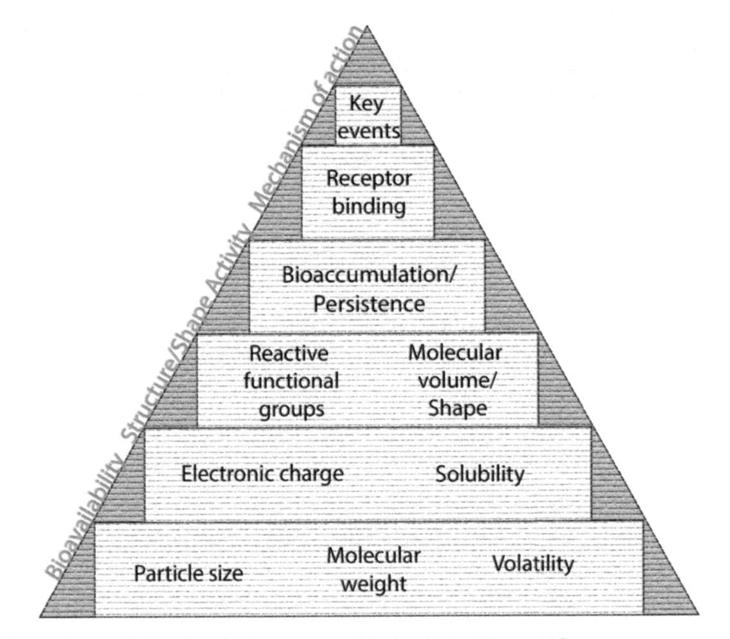


**Figure 7.1** Molecular design pyramid toxicological hazard face.

For example, from the vast amount of available experimental knowledge, we already know that the molecules with $M_W > 400$ Da will not be absorbed in the lungs, those with $M_W > 500$ Da will not be absorbed in the GI tract or transported across the skin. The solubility of the molecule can affect different processes, depending on the medium the molecule is soluble in. If the molecule is hydrophobic, it will tend to float on the surface of the various water bodies (e.g., river, lake, sea) and aggregate with similar molecules, but at the same time, it will easily traverse the double lipid layer of the cell membrane and increase the possibility of accumulating in organisms. On the other hand, if the molecule is hydrophilic, it will be easily distributed through the environment and thus probably more bioavailable, but at the same time more easily excreted from the body. Volatility limits the ability of some chemicals to get into the atmosphere (if vapor pressure is $<10^{-3}$ torr), the tendency to get inhaled ($<10^{-6}$ torr), or reach its lower flammability limit (if $<10^{-8}$ torr). The electronic charge will influence the possibility of a molecule to bind to different particles in the biosphere (e.g., humic acids in humus) or cross various biological membranes.

As more complex processes are governed by the lower-tier "blocks" of the molecular design pyramid, in the following sections we will discuss some specific suggestions that should be considered when designing new chemicals.

## 7.2 How to Design to Reduce Toxicity?

Aquatic toxicity is often the result of either specific interactions of molecules causing adverse effects, or by nonspecific mechanisms.[1] Most chemicals such as alcohols, ethers, weak organic acids and bases, chlorinated hydrocarbons, ketones, and some nitro compounds, among others, cause direct damage to the cells of aquatic species by disrupting the normal cellular functions. On the other hand, some molecules will cause adverse effects by exhibiting specific interactions via inhibition of some enzymes, electrophilic attack on macromolecules such as DNA or RNA, interruption of the CNS functions, blocking of the energy processes in the cell, etc.

Successful molecular design strategies to reduce aquatic toxicity that has been identified include the modification of physicochemical properties of molecules.[2] As the increased molecular weight and size reduce the bioavailability toxicity will also decrease, thus increasing $M_W$ over 1000 Da can be one of the successful strategies as the bioavailability becomes nearly negligible.[3] However, the possibility of forming smaller toxic products from the molecular breakdown should be taken into consideration as well. Another important factor is the octanol-water partition coefficient, i.e., log $K_{ow}$, which governs lipophilicity and water solubility. With the increased lipophilicity the toxicity of the nonionic organic compounds rises until the value of ca. 5. When the log $K_{ow}$ is > 5, acute toxicity and bioavailability decrease, but the bioaccumulation potential rises. If the log $K_{ow}$ is < 1 then the toxicity becomes minimal.[2a] Additionally, lower water solubility means lower bioavailability and at the same time lower toxicity. Reduced electrophilicity of molecules is related to the reduced toxicity, which is correlated with the LUMO (lowest unoccupied molecular orbital) energy, i.e., the higher the LUMO energy is the weaker electrophile the molecule will probably be. Compounds with values of LUMO energies > 2 eV were shown to have reduced aquatic toxicity.[1] Of course, the strategies listed above can be used for the reduction of toxicity in other organisms as the principles governing the exhibition of adverse effects are similar.

If designing to reduce human toxicity one can avoid using, or otherwise modify, problematic pharmacophores. For example, an easily oxidizable C-H bond can be introduced (e.g., toluene instead of benzene); the leaving group potential should be reduced (e.g., bromide can be changed with chloride); furans can be methylated and chlorinated; a number of fused aromatic rings in polyaromatic hydrocarbons (PAHs) should be decreased, or fluorine substituents can be introduced; linear alkanes should be substituted (e.g., hexane to 2,5-dimethyl hexane); benzoquinone can be modified by adding fused aromatic rings to it.

## 7.3 How to Design for Minimized Ignition?

When considering the ways to prevent or minimize the ignition potential of a chemical, it is first important to understand what factors are the crucial contributors to the ignition. As mentioned earlier, and as illustrated in Fig. 7.2, there are three components necessary for the ignition to happen—fuel, a source of heat, and an oxidizing agent. Taking out one corner of the ignition triangle will eliminate (or at least reduce) the ignition potential. Each of these components can be independent of the inherent properties of the chemical, such as the oxidizer that is usually provided by the environmental oxygen, or the sources of heat that are mostly external. However, a lot of chemicals can provide their own source of oxidizer or can decompose providing enough heat necessary for ignition. Additionally, gases and liquids or solids that can volatilize provide the fuel corner of the triangle.

Parameters that are frequently used to characterize the flammability of a chemical substance are lower and upper flammability level, flash point, limiting oxygen concentration, and autoignition temperature.[4] Thus, potential manipulation of these parameters could be used for the design of the chemicals with lower ignition potential. Flammability levels are influenced by the pressure and temperature and a wider range represents a higher risk of ignition. This is also directly related to the vapor pressure of a molecule indicating how likely it is that a chemical will produce flammable vapors. A flash point refers to the minimum temperature at which an ignitable mixture of gases exists above a liquid surface but can also be applied to some solids such as naphthalene or camphor that sublimate when heated. As we mentioned above, some chemicals contain their own source of oxidizers that contribute to the ignition potentials. For example, peroxide decomposition can reduce the ignition energy of a chemical by providing a sufficient amount of oxygen. The potential strategy for reducing ignition is then to avoid the strong oxidizing groups such as peroxide, permanganate, halogen or nitrogen oxides, etc. Sometimes, chemicals can undergo transformations that will produce the heat necessary for the ignition. The polymerization of some monomer chemicals can generate enough heat to result in ignition. Thus, some ways to reduce ignition energy of these chemicals can be the use of additives or functional groups, i.e., polymerization inhibitors that will prevent a polymerization reaction from occurring; incorporating antioxidants that will prevent an oxidation reaction; or introducing stabilizers that will prevent the decomposition of a product.

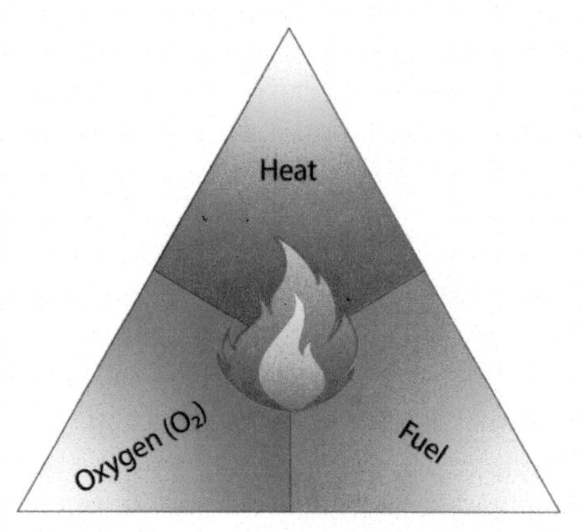


**Figure 7.2** Ignition triangle.

Another valid strategy to minimize ignition potential, especially in the case of materials is to encourage the formation of char on the surface that will prevent the propagation of the flame and further degradation of the substance. This can be achieved by applying the coatings on the surface of the material, which is one of the widely used methods for flame-proofing. However, a large number of these coatings are based on the molecules that, while possessing fantastic flame-retardant properties, act as environmental and even health hazards. Decabromodiphenyl ether (decaBDE) is a molecule used as an additive to flame-retardant substances used throughout the world. As such, it is not a part of the polymer molecule and can leave the material under specific conditions. DecaBDE and other PBDEs in flame-retardant coatings extend the time between the ignition and full combustion which gives enough time for people to evacuate burning structures or extinguish smaller scope fires. However, it has become evident that this molecule, along with other PBDEs persist in the environment, bioaccumulate, and are known human health hazards due to the interference with the endocrine system and neurological development.[5] Meerts et al.[6] hypothesized that the PBDEs interfere with the thyroid hormone metabolism due to the structural similarity to the T4 thyroid hormone. PBDEs competitively bind to the transthyretin interrupting the normal cell metabolism. Exposure occurs through ingestion, inhalation, and dermal adsorption of dust particles originating from the flame-retardant coatings of electronic components,[7] furniture cushions, mattresses, upholstery textiles, carpet backings, aircraft, vehicles, and building materials.[8]

Additionally, PBDEs are structurally similar to other chemicals of concern such as polychlorinated biphenyls and other halogenated aromatic hydrocarbons which make them especially troublesome since they are not easily degraded or metabolized. Thus, a lot of effort was invested in researching alternative flame retardants. Grunlan's group developed water-based technologies based on environmentally relatively benign processes for layer-by-layer (LbL) deposition of flame-retardant nanocoatings on textile materials, and for using aqueous polyelectrolyte complexes (PECs) consisting of polyvinylamine and poly(sodium phosphate) mixed with polylactic acid in the production of self-extinguishing textiles (cotton and nylon-cotton blend), and self-extinguishing filaments used in the 3D printing.[9] Here, polyphosphate promotes the formation of char that will protect the coated surface from the flame. Other technologies rely on the use of biomacromolecules deposited in layers on the textiles to protect the fabric from flames.[10] The biomacromolecules are much more advantageous as they are environmentally friendly and should not generate dangerous decomposition products.

## 7.4 How to Design to Minimize Explosivity?

One of the important concerns while working in the laboratory is safety when dealing with any chemical compounds. Certain chemicals possess the potential to cause an explosion due to the presence of certain types of functional groups as we mentioned earlier. Thus, when considering the safer design of chemicals, it is important to exclude the typical functional moieties that could contribute to the explosivity if possible.[11] These are: acetylene (–C=C–), azo (–N=N–), diazo (=N=N), organic/metal azide (R/M–$N_3$), diazonium salts (R–$N_2^+$), fulminate (–C=N–O), halogen amine (=N–X), nitrate (–$ONO_2$), nitro (–$NO_2$), aromatic or aliphatic nitroamine (=N–$NO_2$ or –NH–$NO_2$), nitrite (–ONO), nitroso (–NO), ozonide (–$O_3^-$), peracids (–CO–O–O–H), peroxide (–O–O–), hydroperoxide (–O–O–H), metal peroxide (M–O–O–M), nitrogen metal salts (=N–M). Additionally, a number of halogen–oxysalts such as bromate ($BrO_3^-$), chlorate ($ClO_3^-$), chlorite ($ClO_2^-$), perchlorate ($ClO_4^-$), iodate ($IO_3^-$), are potent contributors to explosivity.

Another consideration that has to be taken into account when designing molecules to minimize explosivity is the potential of chemicals to form explosive levels of peroxides, either by their inherent reactivity, upon concentration via evaporation or distillation, or by autopolymerization. Chemicals such as butadiene, isopropyl ether, and vinylidene chloride can form explosive levels of peroxides even without concentration; acetal, acetaldehyde, cumene, cyclohexane, dioxanes, methylcyclopentane, vinyl ethers and other secondary alcohols can form dangerous levels of peroxides when concentrated; while chemicals such as acrylonitrile, chloroprene, styrene vinyl acetate, or vinylidene chloride can autopolymerize as a result of peroxide formation. Thus, the design guidelines should carefully consider the similarities to peroxidizable chemicals.

## 7.5 How to Design to Reduce Absorption?

The molecular structure is an important factor influencing the lipophilic character of a chemical. There are a number of different methods developed for the prediction of molecular log P based on the whole molecules (where only molecular parameters such as polarizability, size, and H-bond acceptor strength is considered), fragments (functional groups), atoms, and following constructionist and reductionist principles.[12] The group contribution approach is conceptually simple and gives fast and relatively accurate estimation for most organic compounds. However, correction factors have been used to account for various interactions of the groups (e.g., hydrogen bonds, non-covalent interactions). The quantitative structure–property relationship (QSPR) method has been widely used for the modelling of most ADME properties.

Absorption is mostly affected by molecular volume, the dipolar character of the molecule, and acidity/basicity of the hydrogen bond,[13] thus, these structural properties can be modified to adjust log P values. Molecular volume represents the size of the cavity formed when the molecule is solubilized by the solvent, and it is directly related to the molecular weight; dipolar nature of the molecule plays the role with the orientation of the molecule with the solvent; hydrogen bonding with the solvent is dictated by the hydrogen bond accepting nature (basicity) and donating nature (acidity) of the molecule. These properties have been used for the prediction of partitioning behavior and the assessment of the various assay methods. However, the lipophilic character of a molecule, and the ability to predict it, changes with the pH and presence of the buffer, which affect the polarity of the aqueous phase; the molecular properties of the phases, leading to a different interaction between solvents and solute; and the presence of other solvents and solutes in the mixture, that can interact with molecules and change their partitioning behavior.

While considering the modifications to the molecules to affect its adsorption, it must be considered that all physicochemical properties are affecting each other. That means modifying a group of properties resulting in increased water solubility will reduce lipophilicity, and vice versa. The structure can be modified by reducing the size of aliphatic chains or increasing the branching which will reduce the lipophilicity of the molecule. Water solubility can be increased by introducing ionizable groups such as amino or carboxylic groups, which is one of the most effective changes to increase hydrophilicity. The addition of $-OH$ and $-NH_2$ groups, i.e., H-bond donors/acceptors, at different positions in the molecule can dramatically change the water solubility. Additionally, the use of polar groups along with the reduction in molecular weight is another effective approach.

To reduce the absorption of chemicals through main routes of exposure to the body (respiratory, dermal, and gastrointestinal), molecules can be designed following several useful strategies. Some of the guides rely on Lipinski's rule of five that predict which molecules will exhibit poor permeation or absorption. Respiratory absorption through lungs can be reduced by increasing the molecular weight to over 400 Da and size to over 5 μm. Further, the vapor pressure of the molecule should be below $<10^{-6}$ torr, while the blood-to-gas partition coefficient should be <1. The absorption through the skin can be limited by increasing the water solubility of the molecule (e.g., by increasing the polarity), increasing the molecular weight over 400 Da promoting the solid state of the chemical, and lowering the log P values. Finally, the absorption through the GI tract can be reduced by increasing the molecular weight over 500 Da, increasing the particle size over 100 nm, keeping the log P below 0 or over 5, introducing groups that will promote ionization at pH < 2, increasing melting point > 150 °C (molecule is solid at body temperature), avoiding the use of hydrolysable bonds in the molecule (hydrolysable ester and amide bonding), and increasing the number of H-bond donors and acceptors (more than 5 H-bond donors, more than 10 H-bond acceptors).

## 7.6 How Can We Change Properties/Structure to Minimize Inhalation?

During the breathing process, various molecules can be inhaled into the lungs in form of a gas, aerosol, or solid particle. The properties guiding this are related to the physicochemical properties such as size (or mass), solubility, boiling point, charge, etc. Larger molecules or particles are usually filtered in the nose, esophagus, or finally in the mucous layer of the respiratory tract. Only much smaller molecules of a diameter smaller than 3–5 μm can penetrate deep into the lungs. Most of the substances in gaseous form are easily inhaled as they are usually very small and very volatile (Fig. 7.3).

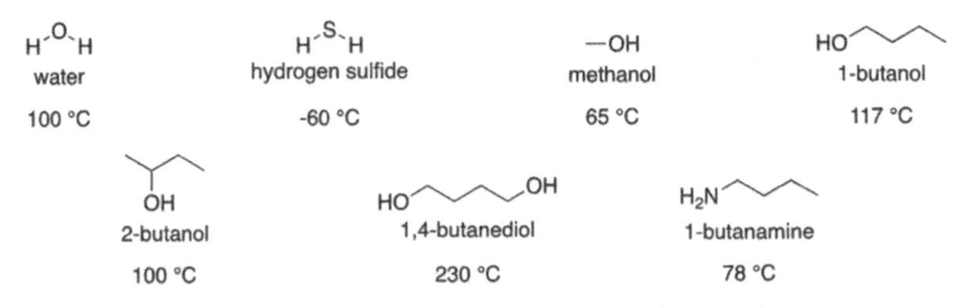

**Figure 7.3** Boiling points of selected compounds.[14]

Therapeutics developed for the treatment of lung diseases have demonstrated increased efficiency when the particle size was smaller, i.e., by permitting the deeper penetration and deposition of the drug in the lungs.[15] Thus, the inhalation potential can be reduced by increasing the size of the molecule or by improving the tendency of the molecules to self-aggregate.

Otherwise, structural modifications that can reduce volatility could be pursued as well. Molecules forming strong intermolecular forces will be harder to separate, decreasing their volatility and vapor pressure. For example, hydrogen sulfide is more volatile than water, since the $H_2S$ molecules are bound via dipole-dipole interactions, compared to $H_2O$ molecules bound by much stronger hydrogen bonds. Another example is the lower boiling point of methanol compared to butanol. While both alcohols have a polar OH group that can form hydrogen bonds, butanol has a longer hydrocarbon chain that forms stacks via London dispersion forces.

## 7.7 How Can We Change Properties/Structure to Minimize Transport Across the Gastrointestinal Tract?

We have already described the complexity of the GI tract earlier, where different areas are responsible for the transport, digestion, and absorption of different types of molecules. For example, organic acids and bases tend to be absorbed by passive diffusion in the GI tract where the pH will favor their most nonionized form, partially because of the large range of pH values in the gut. Naturally, this complicates the intentional design for reduced transport and absorption across the GI tract, since molecules can get hydrolyzed by stomach acids or biotransformed by enzymes.

The absorption of a xenobiotic and transport through the membranes in the GI tract generally depends on its lipid solubility and size.[16] An increase in lipophilicity will typically increase the absorption rate, while the increase in size reduces the transport. However, molecules can also interact with different transport molecules, such as carrier and channel proteins. For example, small hydrophilic molecules are transported easily through cell membranes via transmembrane proteins.[17]

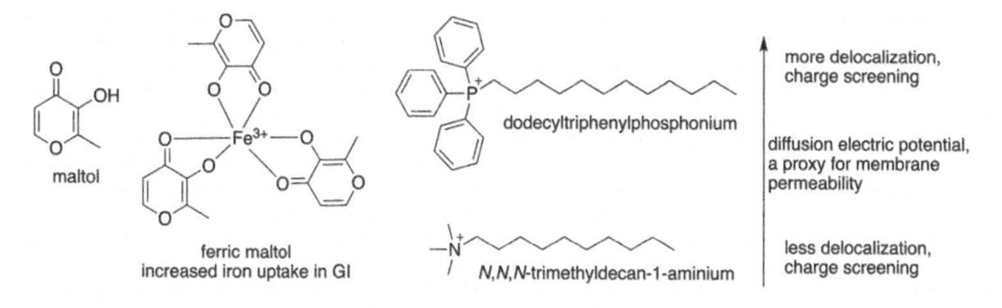


**Figure 7.4** Iron chelation and effects of charge delocalization and screening on permeability.[18]

Most of the heavy metal ions, such as lead, are usually not absorbed in the GI tract. However, chelating agents can form complexes with these ions increasing their lipophilicity (Fig. 7.4). Thus, chelation should be avoided where possible.

Additionally, permanently charged molecules vary in the rate of permeation depending on their ability to spread the charge, such as distribution over several aromatic rings.[19] The more delocalized the charge is, the higher the passive membrane permeability will be (Fig. 7.4).[18b]

## 7.8 How Can We Change Properties/Structure to Minimize Transport Across Skin?

As the largest organ of the body, the skin is the first line of defense against a large number of xenobiotics. As mentioned earlier, dermal layers are practically impenetrable for a number of aqueous-soluble molecules and most ions. On the other hand, molecules with increased lipophilicity can more easily get transported through or between the cells, unless they are highly lipophilic and accumulate in the cells. Thus, amphiphilic molecules possess the highest potential for transport across the skin and into the body.

Minimizing the transport across the skin can be related to Lipinski's rule of five as mentioned earlier. One of the strategies for minimizing the skin transport would then revolve around the modification of the log P values. While the RO5 states molecules with log P > 5 would have difficulty permeating the epidermis, it was also observed that molecules with log P < −1 have difficulty passing through the top keratinous layer, while those having a log P value of 1 to 3 have been shown to exhibit proper transdermal permeation.[20]

Another factor influencing the permeation across the skin is the size (or $M_W$) of the molecule. Smaller molecules are more likely to get transported by passive diffusion than larger molecules. RO5 states that molecules with $M_W$ > 500 Da would have difficulty being absorbed, however, some molecules larger than that can also be absorbed.[21]

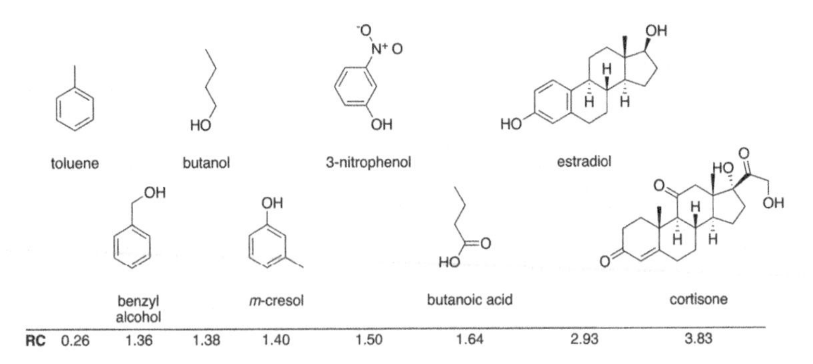


**Figure 7.5** Membrane penetration retardation coefficients for selected compounds.[22]

Since the skin cells consist of numerous types of molecules (e.g., sugars, proteins, enzymes, lipids, etc.), they can interact with a xenobiotic in various ways, such as by hydrogen bonding or via weak non-covalent interactions. These interactions will affect the rate of the transportation of molecules across the skin. For example, the keratinous layer acts mainly as a hydrogen bond acceptor, so an increase in the number of hydrogen bond donor groups such as OH or NH, could inhibit the permeation of a molecule through the skin.[22] Figure 7.5 lists membrane penetration retardation coefficients for some molecules that can come into contact with the living organisms.

## 7.9 How to Design to Modify Active and Passive Transport?

As the passive transport is determined by the lipophilic character of a chemical, manipulation of the log P parameter, will influence the level of absorption. Additionally, we mentioned that the size of the chemical can determine the type of transport that is preferred for the absorption, thus, changing the size of the molecule will influence the rate of the said transport. For example, if the molecular weight is larger than 300 Da then it will not be absorbed in the GI tract; if it is larger than 500 Da it will not get absorbed in the lungs; if it is larger than 1000 Da it will not get absorbed through the skin.[1] The size of the molecule can be increased by introducing branching or sterically larger functional groups and increasing the number of freely rotatable bonds, of course taking into consideration the other consequences this would cause. Another important factor that influences the transport of the chemical is its ionic character. As some molecules exist in the different ionic states depending on the pH of the environment, this aspect can also be used to manipulate the rate of transport. Decreasing the degree of ionization under physiological conditions will generally improve the transport across the membranes, while the introduction of polarizable groups under the relevant pH (e.g., at the skin surface, in the stomach or intestines, inside the lungs).[1] Reducing the number of H-bond donors and acceptors will improve transport, while the increase in their number will reduce transport.

An example of efficient structural modification for tailored active transport for reduced toxicity is the case of Paraquat and Diquat (Fig. 7.6). Paraquat is a nonselective contact herbicide developed in the 1960s that causes pulmonary toxicity in humans and some animals. The mechanism of action involves active transport into alveolar cells via the polyamine transport (PAT) system.[1,23] Once the Paraquat molecule gets into the cell, it binds to DNA and RNA electrostatically due to its positive charge. On the other hand, Diquat possesses the same efficiency but exhibits much less toxicity to mammals. The reason is the reduced retention due to the increased steric hindrance of the Diquat structure which lowers the affinity towards the amines in the PAT system.

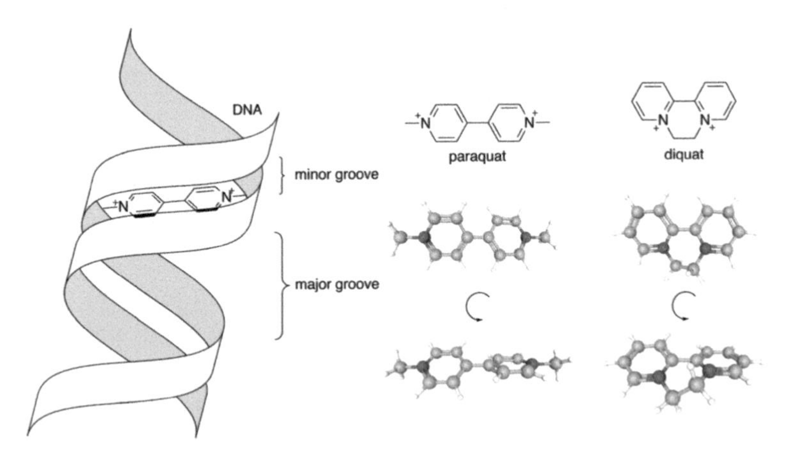


**Figure 7.6** (Left) DNA double helix with intercalated paraquat molecule. (Right) Paraquat and diquat molecules with the 3D molecular structures.[23]

## 7.10 How to Design to Reduce Persistence?

As the persistence of chemical compounds in the environment can cause a range of adverse effects due to their inherent stability and potential to interact with various biological systems, it is very important to consider the ways to promote the degradation of these chemicals at the end of their useful life. The persistence of chemicals can be reduced by reducing their half-life by promoting degradation. Different strategies can be employed to promote various types of degradation mentioned in earlier chapters. From the molecular design perspective, this can be done by modifying the structure and physicochemical properties of molecules. In the following sections, we will describe specific molecular design suggestions to promote different types of degradations discussed in the previous chapters.

### 7.10.1 Biodegradation

The strategy for promoting biodegradation relies on making the molecules more bioavailable to the microbial communities that will use them for growth or generating cellular energy. Since the bacteria can degrade chemicals both in the presence and absence of oxygen, we can distinguish between two sets of suggestions, i.e., aerobic (Fig. 7.7) and anaerobic degradation (Fig. 7.8).

Aerobic bacteria contain a set of enzymes that degrade molecules in the presence of oxygen, thus the inclusion of functional groups that are susceptible to enzymatic hydrolysis such as esters and amides will generally increase the biodegradability. As we mentioned earlier, one of the key steps in the metabolism of many chemicals is the insertion of molecular oxygen into their chemical structure by oxygenase enzymes. Phenyl rings and unsubstituted linear alkyl chains that have more than four carbons are particularly susceptible to these insertions. Additionally, it was observed that the small molecules that already have oxygen in the structure biodegrade more easily than structurally similar molecules that do not have it. Aliphatic hydrocarbons will degrade slower than their related alcohols and carboxylic acids,[24] while cyclohexanol and cyclohexanone degrade easier than cyclohexane.[25]

On the other hand, molecules will exhibit increased resistance to biodegradation if they have halogen atoms in the structure, especially if more than three halogens are present.[26] Additionally, even if the phenyl rings are present in the structure, electron-withdrawing substituents on the rings will hinder the biodegradability of the molecule. If more than three aromatic rings are fused in the structure, such as the case in the number of PAHs, the molecules will generally be more persistent. The increase in size is considered

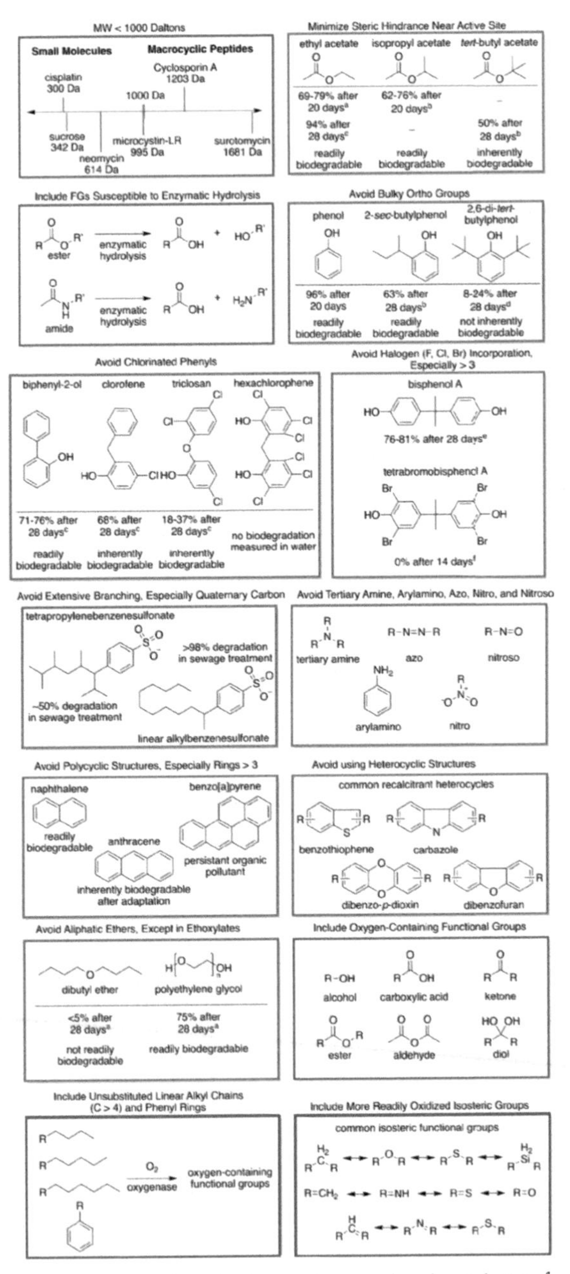


**Figure 7.7** Design suggestions for improving aerobic degradation.[1]

---

[1](a) Standard methods for the examination of water and wastewater 1971, 13th ed., American Public Health Assoc, NY, and modified BOD test; (b) OECD Guideline 301 D; (c) OECD Guideline 301 B; (d) OECD Guideline 302 C; (e) OECD Guideline 301 F; (f) OECD Guideline 301 C.

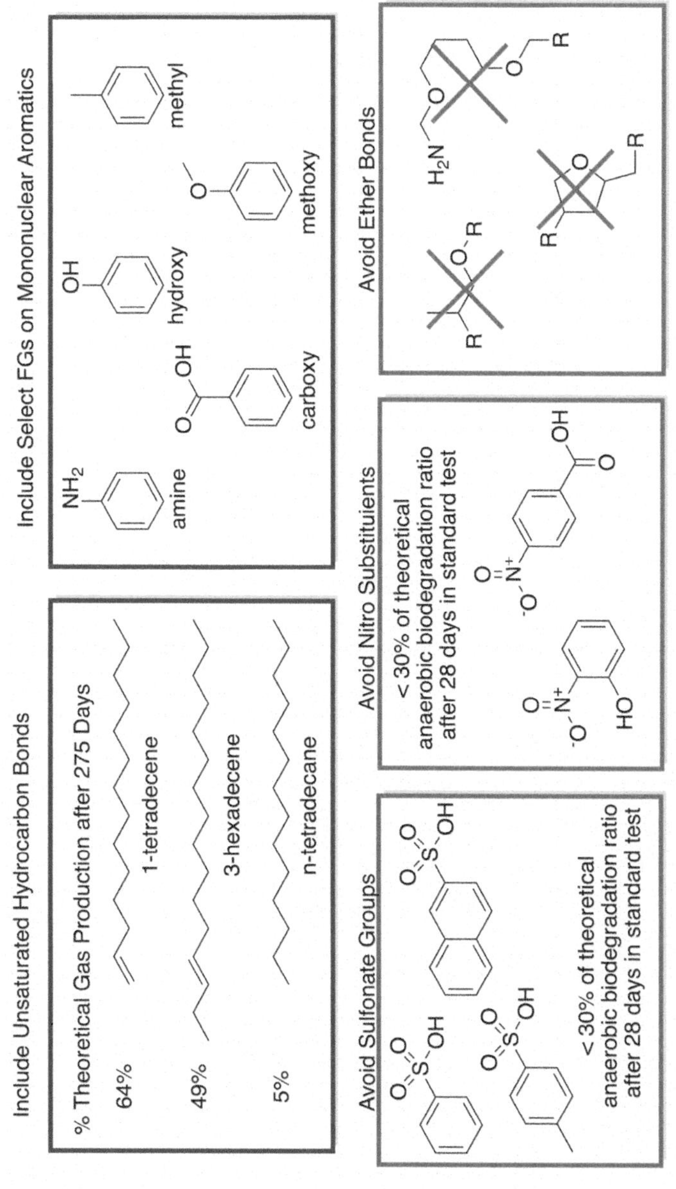


**Figure 7.8** Design suggestions for improving anaerobic degradation.

to hinder biodegradability since the enzymes that are involved in the process cannot access the molecule. Thus, extensive chain branching especially if quaternary carbon is present mostly inhibit biodegradation, but of course exceptions exist (e.g., cholesterol, pentaerythritol, vitamin A).[27] More design suggestions, along with some real-life examples are shown in Fig. 7.7.

Similar to aerobic degradation, decreasing the molecular weight and steric hindrance at the active site of the molecule will increase its anaerobic biodegradation. Also, the presence of several halogen atoms, extensive chain branching, bulky ortho substituents on phenyl rings, etc. will hinder biodegradation. On the other hand, olefins generally biodegrade faster than respective saturated hydrocarbons; mononuclear aromatic compounds should contain at least one amino, carboxy, hydroxy, methoxy, or methyl substituent; and nitro substituents and ether bonds should be avoided.

### 7.10.2 Photolytic Degradation

Sometimes it can be beneficial to promote photolytic degradation as a strategy to tailor the life of a chemical in the environment. This can be achieved with the inclusion of photoactive, i.e., photolabile moieties. As mentioned in Chapter 3, these moieties have weak bonds whose dissociation energies are less than the absorbed UV radiation, leading to their breakage. Figure 7.9 shows some typical examples of the groups that can be included to promote photolysis.

However, this type of degradation can be problematic since it can lead to the formation of toxic products if the degraded chemical is not fully mineralized. Thus, it should be carefully considered before the photoactive groups are included when designing chemicals.

### 7.10.3 Hydrolytic Degradation

All chemicals that can get into aquatic environments, whether freshwater or marine, can be potentially hazardous. However, this hazard can be mitigated if the chemical can safely degrade when it gets into this environment. Thus, when considering safer design strategies, it becomes highly beneficial to increase the water solubility of the molecule by including polar functional groups for example. Additionally, decreasing the size and molecular weight can

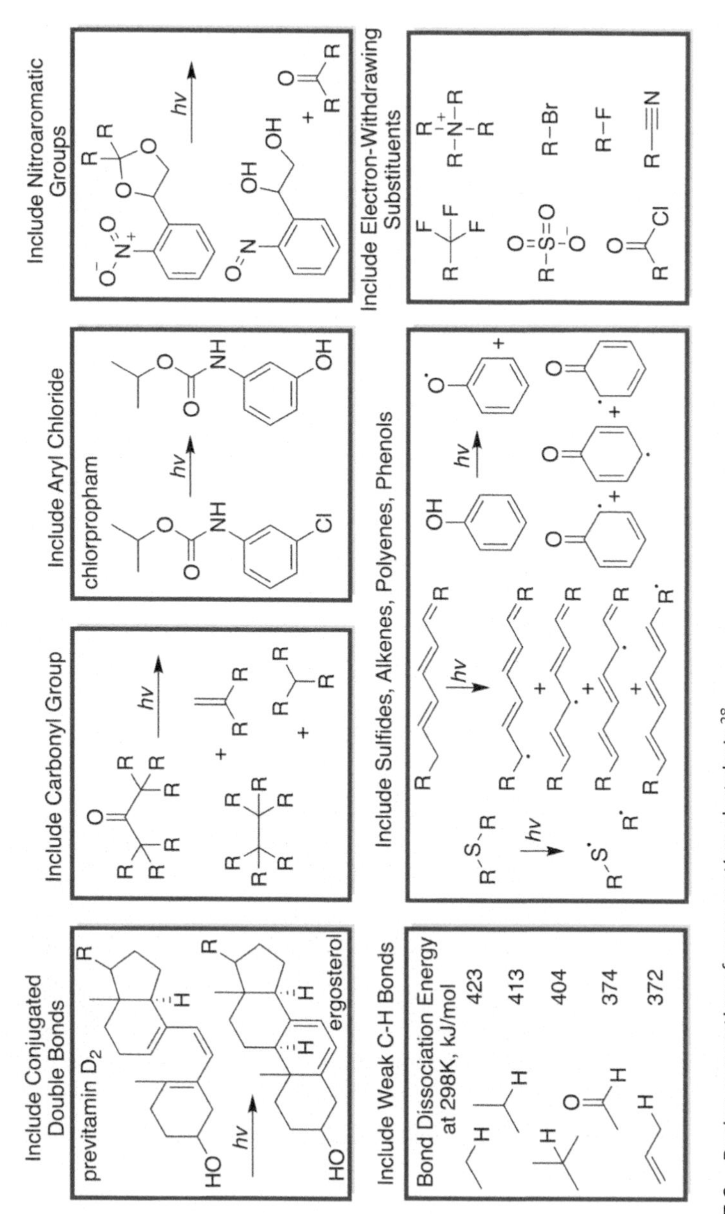

**Figure 7.9** Design suggestions for promoting photolysis.[28]

lead to easier hydrolysis as the molecule becomes more available to water molecules. Further, electron-withdrawing substituents and hydrolysable bonds lead to higher hydrolysis rates (see Fig. 7.10). When designing molecules for increased hydrolytic degradation potential, one can, for example, consider known hydrolysis rates for the sequence: anhydride > ester > amide > ether functional group.

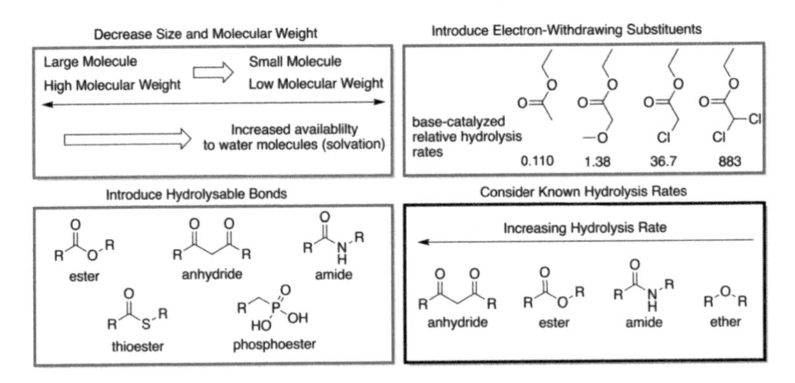


**Figure 7.10** Design suggestions for improved hydrolysis.[29]

## 7.10.4 Thermal Degradation

Chemicals are usually designed to resist thermal degradation, especially so when degradation leads to products that can be very reactive or toxic. However, for some purposes, it can be useful to promote the thermolysis of molecules. Similar to photolysis, the introduction of thermolabile functional groups such as carbonate, increases the thermal degradation rate. This is not only important in the design of safer molecules but the design of greener solvents as well. For example, ionic liquids based on azolate anions are flammable, while those based on imidazolium and pyridinium ions are not.[30] Figure 7.11 shows the trend in bond strength of some chemical bonds and bond dissociation energies that can be considered when deciding how to design molecules for thermal decomposition.

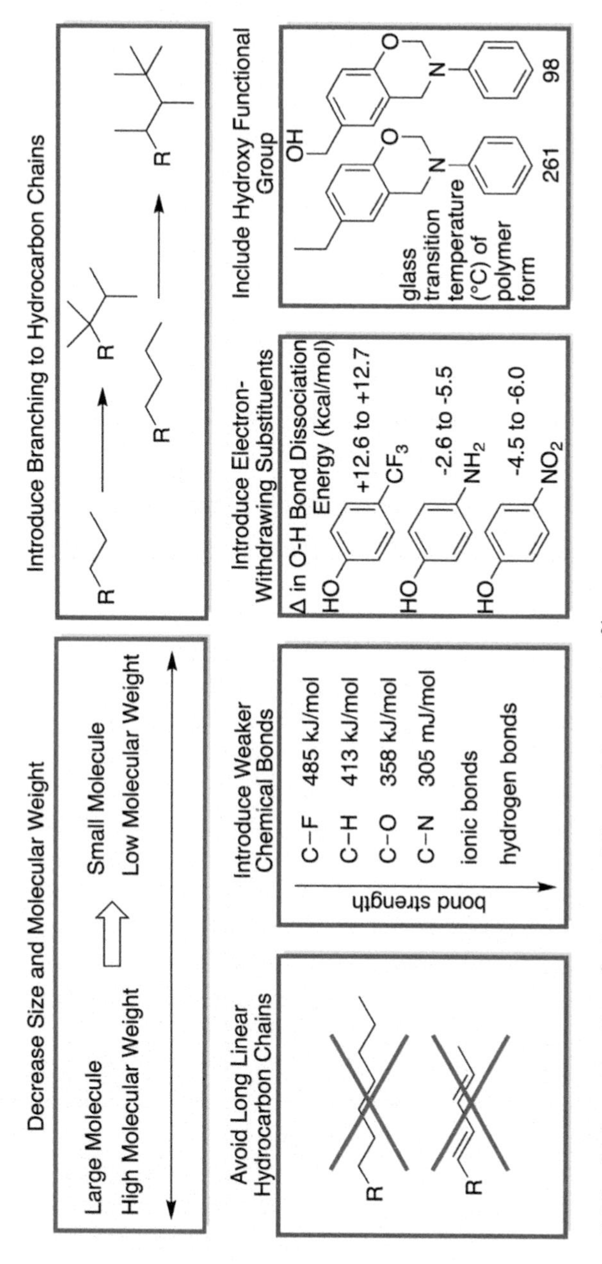


**Figure 7.11** Design suggestions for increased thermal degradation.[31]

If the molecule is designed with thermal degradation in mind, decreasing the size and molecular weight will contribute positively. Long linear hydrocarbon chains are generally to be avoided, as the packing of molecules in bulk will favor non-covalent intermolecular interactions that can increase energy requirements for their decomposition. Thus, the introduction of branching in the hydrocarbon chain should improve thermal degradation. Another valid strategy can be to introduce electron-withdrawing substituents to aromatic rings, as they can decrease electron density to the substituents at ortho- and para-positions, thus making them more susceptible to dissociation. Additionally, as can be seen from Fig. 7.11, the introduction of the OH functional group to the polybenzoxazines reduces the glass transition temperature of the polymer form compared to the one without the OH group.

## 7.11 How to Design to Minimize Bioaccumulation/Bioconcentration?

As mentioned earlier, bioaccumulation and bioconcentration of chemicals pose a huge risk for all organisms across the environment. Bioconcentration refers to the intake and retention of a certain chemical through the respiration process that occurs through the air for terrestrial species, and water through aquatic species. Bioaccumulation on the other hand is the intake and retention of chemicals occurring through any exposure route. Often, highly lipophilic substances such as heavy metals or organic compounds (especially organohalides) get deposited in the fat tissue of different organisms and pass through the food chain to the species that are at the top (e.g., predator species—eagles, bears, humans) where they can reach high, possibly dangerous concentrations, through the biomagnification process. Thus, they should be avoided.

Design suggestions are shown in Fig. 7.12. Chemicals that biomagnify usually have log P values between 3 and 6, and while they pass through the cell's lipid bilayer, they usually cannot be easily excreted and stay in the fatty layer. Thus, when designing for the minimized bioaccumulation log P of the molecule should be below 2, noting that some chemicals with log P larger than 6 can still be metabolized (e.g., $\alpha$-tocopherol—vitamin E[32b]). The size of the

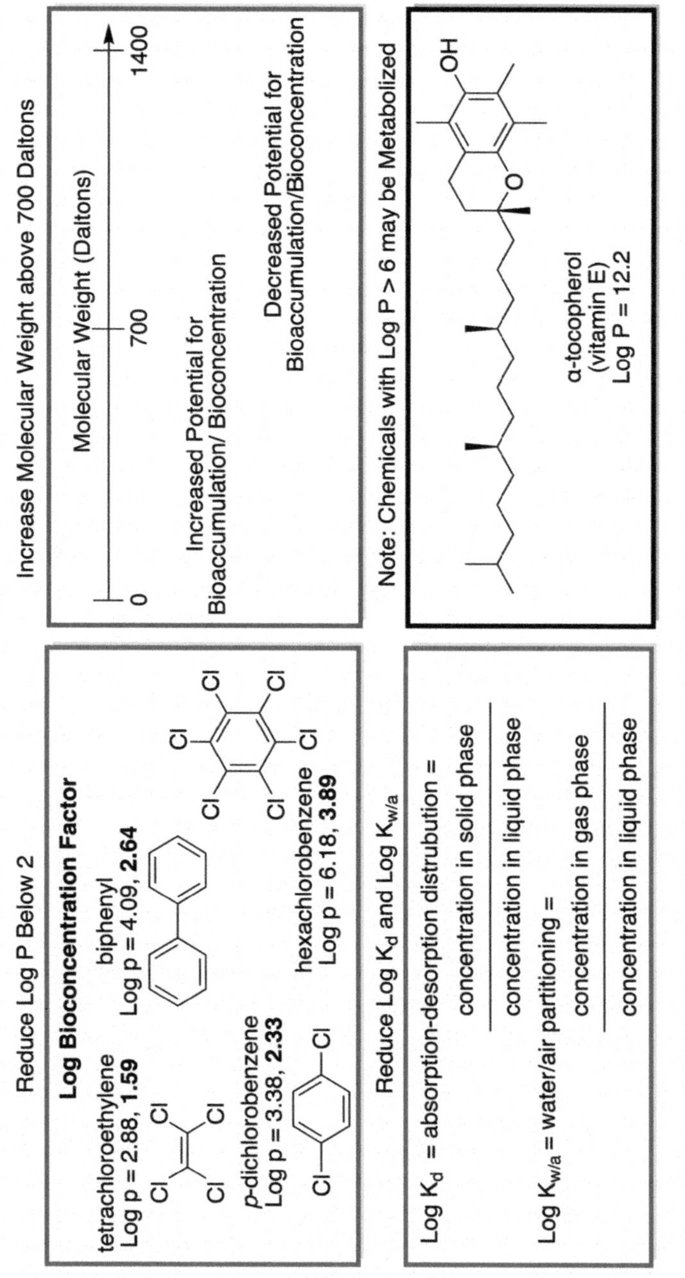

**Figure 7.12** Design suggestions for minimal bioaccumulation/bioconcentration.[32]

molecule is also very important. Molecules will also be less prone to bioaccumulation if the log $K_d$ and log $K_{w/a}$ partition coefficients are reduced. Smaller molecules tend to bioaccumulate easier, while larger molecules ($M_w$>700 Da) have decreased potential towards bioaccumulation.

Some notorious examples of bioaccumulation are naturally occurring mercury and lead bioaccumulation, and the bioaccumulation of man-made halogenated organic compounds.

## 7.12 What Structural Features Can Be Built in to Avoid Bioactivation?

The bioactivation of xenobiotics is a very troublesome mechanism naturally occurring as a part of metabolic processes in the body. Since many functional groups which undergo reactions that are responsible for unwanted bioactivations coincide with the other beneficial transformations it is very difficult to design a molecule in such a way to completely avoid bioactivation.[33] Still, some strategies based on the common bioactivation mechanisms described in the previous section can be taken to avoid bioactivation risk. These strategies include the replacement of the functional groups that possess a high tendency for bioactivation, making the bioactivation less favorable or completely blocked, introducing the sites at the molecule that would be preferred over the sites with 'riskier' functional groups, or reducing electronic density.[34]

One of the most effective ways to reduce bioactivation potential is to replace the functional groups (or fragments) possessing high bioactivation potential, such as aromatic systems containing electron-donating groups, five-membered heterocycles, or phenoxy or chloro-phenyl groups; with substitutes that are resistant to metabolism or can be biotransformed to non-reactive molecules. Some examples used in the pharmacologic practices are shown in Fig. 7.13.

Another successful strategy can be to make the metabolism less favorable by reducing the oxidation potential of the molecule. For example, the replacement of the OH-phenolic group with a fluorine atom has been used to make a compound that does not undergo oxidation in presence of peroxidase.[36] Fluorination is often used as an approach to block the hydrogen radical abstraction in aliphatic molecules, and partially for the systems involving π-bond oxidation of aromatic systems.[34a]

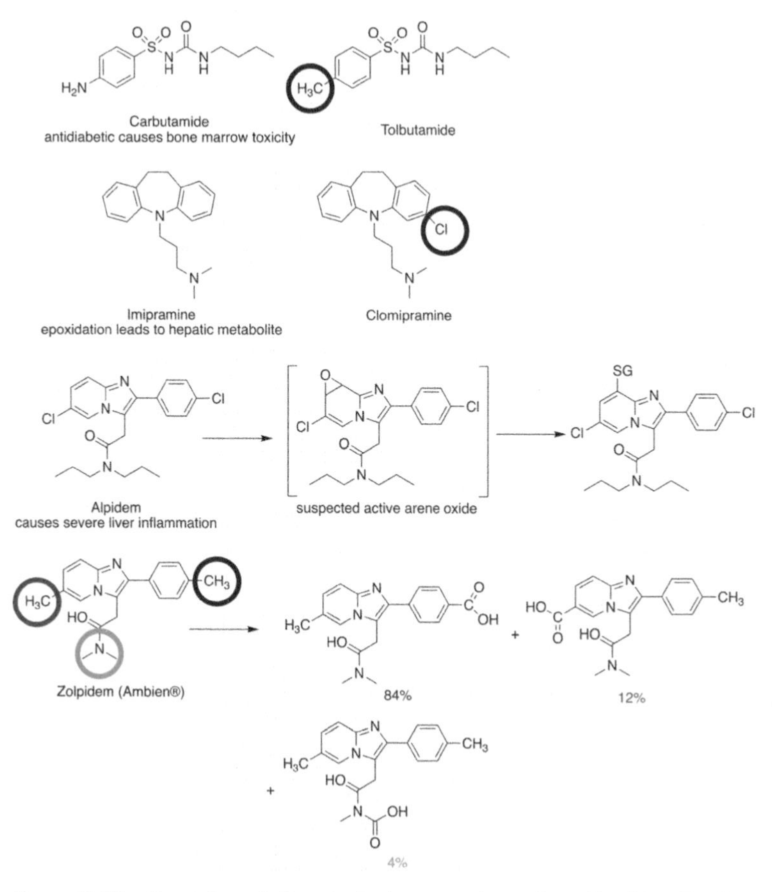


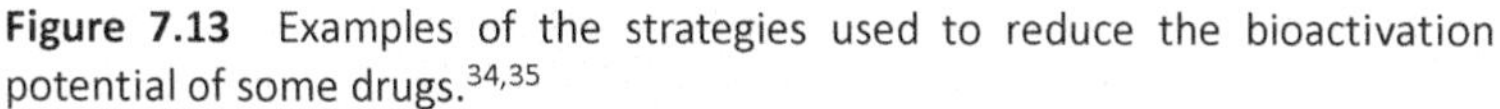

**Figure 7.13** Examples of the strategies used to reduce the bioactivation potential of some drugs.[34,35]

Metabolism mechanisms can also be directed towards a more preferred site of the molecule by introducing functional groups that will be easier targets whose transformation will not result in a harmful molecule. Generally, the presence of some functional groups will usually lead to more benign biotransformations. For example, as shown in Fig. 7.14, esters will hydrolyze to carboxylic acids and alcohols, cyclohexyl groups oxidize to alcohol and ketones, hydroxyl and phenolic groups conjugate to sulfates or glucuronides, etc.

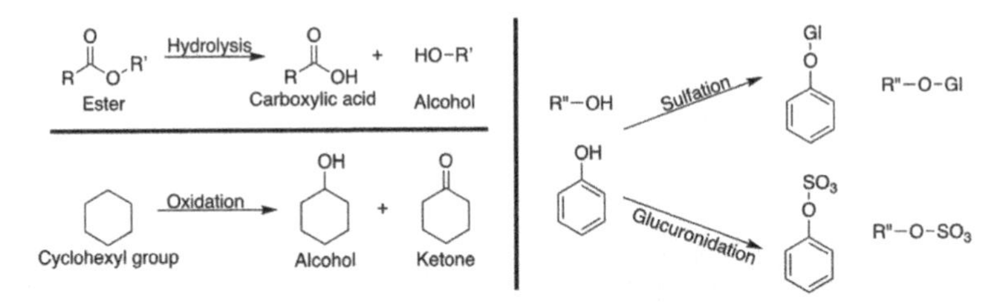


**Figure 7.14** Typical reactions occurring in the metabolism of chemicals.

Manipulation of the electron density can efficiently be used for reducing the bioactivation potential. Some examples include the replacement of the $NO_2$ group in flunitrazepam with bromine resulting in bromazepam or the replacement of bromine with trifluoromethoxy group in halothane anesthetic resulting in desflurane possessing a reduced affinity for bioactivation.[34b]

## 7.13 What Structural Features Can Be Built in to Promote Detoxification?

Generally, the detoxification of harmful molecules will happen naturally. To facilitate this process, it can be beneficial to mimic the classes of molecules already naturally detoxified by the enzymes we mentioned in the previous section following the design suggestions shown in Fig. 7.15. The detoxified compounds should be water-soluble so they can be readily excreted, thus the target molecule should contain functional groups that will be transformed by phase I and phase II reactions to having a more polar character. Coincidentally, this strategy can also take advantage of the modifications that are opposite to those used in pharmacology to increase the stability of the drug candidates.

The design consideration should include incorporation of labile functional groups (e.g., alcohol, amine, ester, ketone, etc.) that can be easily accessible by metabolic enzymes, especially those involved in conjugations reactions from phase II of the metabolism. Cyclization of the molecule should generally be avoided, however, if that is not possible, the ring size should be increased. Also, steric hindrance should be reduced as much as possible, while the number of electron-donating groups should be increased. At the same time, structures, where resonance stabilization of electrophilic metabolites is occurring, should be avoided.

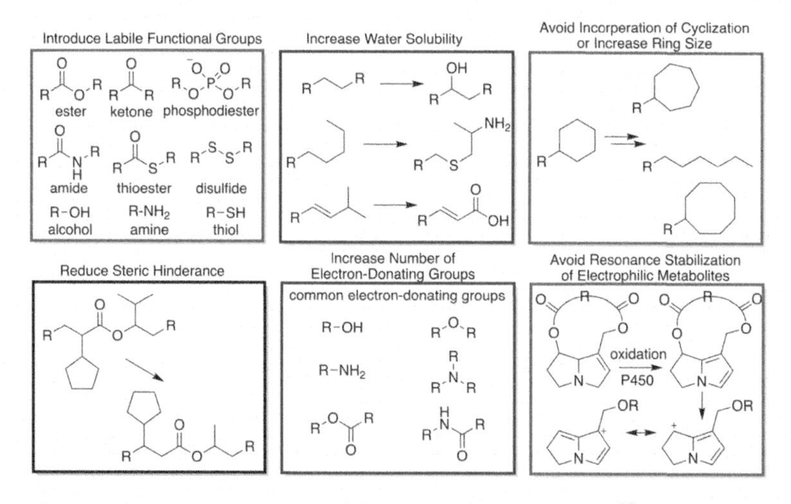


**Figure 7.15** Design suggestions for promoting detoxification.[37]

Of course, it is very important to note that the features that are introduced to promote detoxification can also lead to bioactivation, so the choices should be considered very carefully.

## 7.14 What Properties Promote Distribution?

As we mentioned in the previous sections, the distribution of a chemical depends on several factors that influence the solubility, passage through the cell membranes, and tendency to bind to some proteins existing in the inter- and intracellular space.

The aqueous solubility and lipophilicity of some chemical compounds will have a big influence on their distribution through body fluids and the passage through various biological barriers. Lipophilicity generally influences the passage of molecules, i.e., permeability, through biological barriers such as the blood-brain barrier, consisting of a network of thin-walled capillaries. Having higher lipophilicity will allow the molecule to pass through the barriers without too much difficulty, while higher aqueous solubility means that the molecule will increasingly tend to remain in the extracellular space or blood.

Aqueous solubility will increase if there are ionizable groups present, such as amino or carboxylic groups; reducing the lipophilic character, i.e., the log P value, will result in increased solubility and overall systemic exposure (Fig. 7.16). Aqueous solubility can also be increased by the addition of polar groups, or hydrogen bond donors and acceptors, such as –OH and –$NH_2$ groups. The size of a molecule influences the solubility as well—smaller molecules with lower $M_w$ are generally more soluble and metabolically stable than larger molecules. Permeability of the molecule is increased with the addition of the more lipophilic groups, e.g., longer nonpolar hydrocarbon chains; by esterification of carboxylic acid; replacing the ionizable to non-ionizable groups; reducing the size of the molecule; or by reducing the hydrogen bonding.

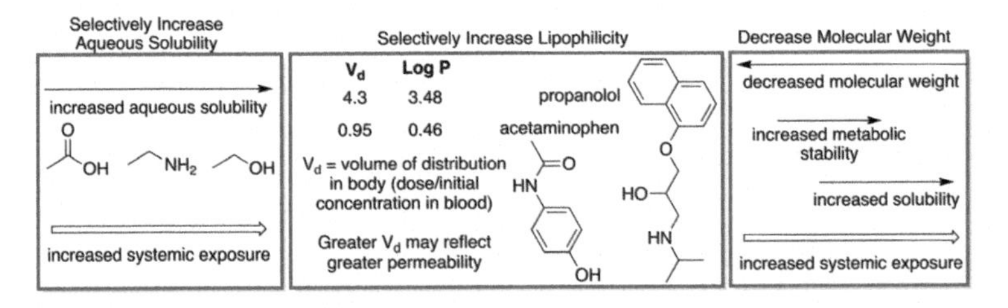


**Figure 7.16** Design strategies for promoting distribution.[38]

The physicochemical properties influencing the solubility and permeability are intercorrelated, and often structural changes that increase aqueous solubility will reduce permeability. This means that the molecules with properly balanced properties will have the highest distribution rate throughout the whole body.

## 7.15 How to Design to Promote Excretion?

Earlier, we mentioned that the metabolic processes in the body naturally transform harmful xenobiotics to be easily transported to the sites where they can get excreted via exhaled air, feces, or urine. The properties that will promote excretion will depend on the route of excretion and can be adjusted following the design suggestions shown in Fig. 7.17.

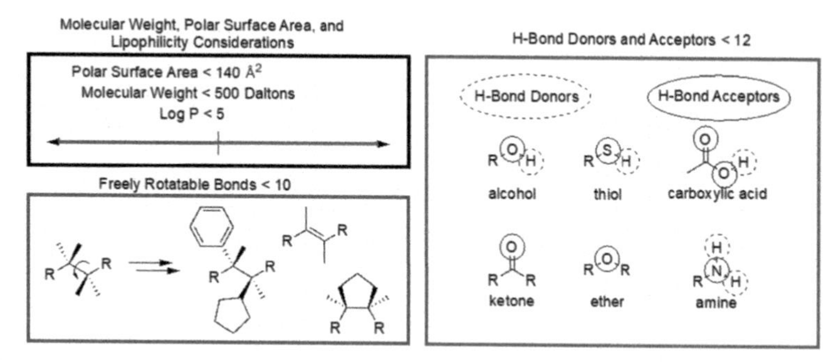


**Figure 7.17** Design strategies for better excretion.

Most water-soluble molecules get excreted through the urine, so the increased aqueous solubility will result in faster removal of these compounds. As for the increased distribution, this can be achieved by adding ionizable and polar groups, reducing the log P and $M_W$, or adding hydrogen bond donors and acceptors.

Lipophilic compounds that cannot be converted to water-soluble molecules usually get excreted with bile via feces by passive diffusion and active transport. Influencing the log P values by adding lipophilic groups, modifying molecular volume, or replacing the polar groups will contribute to the increased excretion rate.

To promote excretion, properties governing the absorptions will also play a role. This includes expanded Lipinski's rule of five suggestions with keeping $M_W$ below 500 Da, log P below 5, having less than 10 freely rotatable bonds while keeping the number of H-bond donors and acceptors less than 12, and reducing the polar surface area to less than 140 Å$^2$.[39]

## 7.16 What Structure/Property Changes Can Be Made to Disfavor Toxic Mechanisms of Action?

Understanding the toxic mechanism of action for some molecule and how it interacts with the receptor are the most important factors in designing for reduced toxicity. This information can be obtained by observing the mechanism of action for a series of similar molecules already documented in the literature or by relying on modelling by various computational methods.

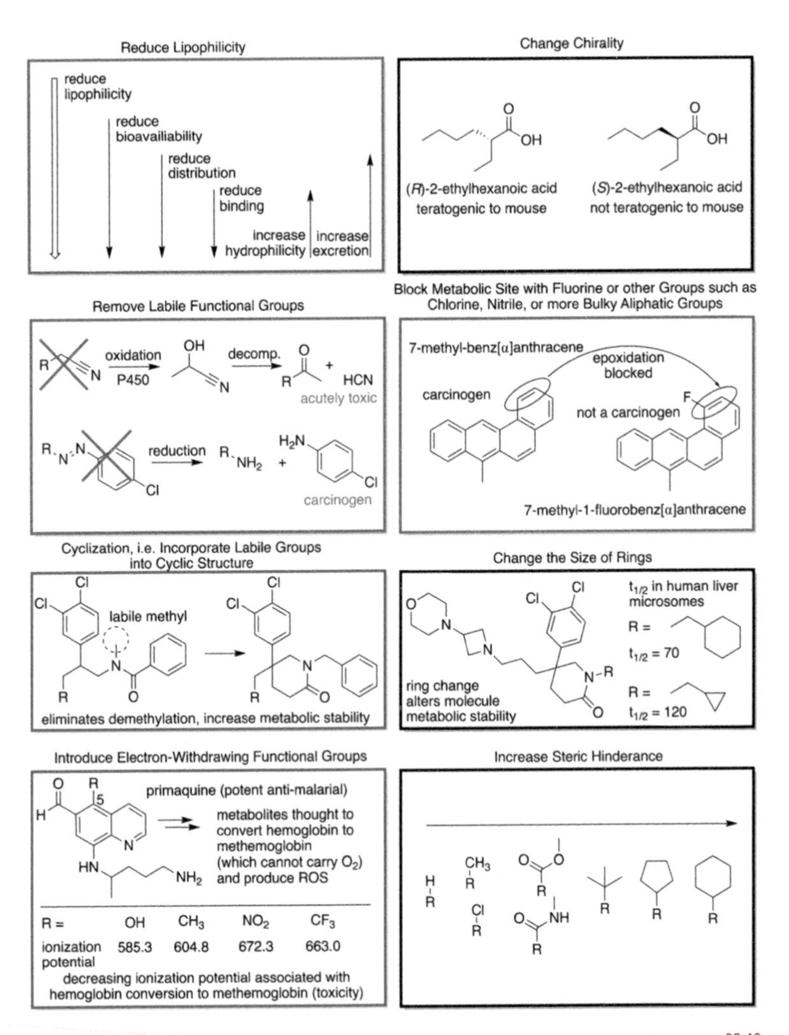


**Figure 7.18** Design strategies for reducing the toxic mechanism of action.[38,40]

Often, the cause of xenobiotic toxicity is metabolic bioactivation. These mechanisms can transform molecules into reactive metabolites or intermediates that can cause adverse effects. The metabolite could be an electrophile that will covalently bind to proteins or DNA, causing mutations that can lead to cancer, impaired function, or immune response. For example, acetamides get converted to radicals that cause oxidative stress; unsaturated bonds can be converted by CYP450 into epoxides that are very reactive and can attach to proteins or DNA; nitroaromatics are reduced to aromatic N-oxide, nitroso, nitro, and nitroxyl radicals that can induce oxidative stress.[38]

To avoid the bioactivation and toxicity of molecules, certain structural modifications can be made to improve their stability. These modifications are intended to affect the binding to the metabolic enzymes and change the reactivity of the molecule at the labile site. Some design suggestions are shown in Fig. 7.18. For example, reducing lipophilicity will prevent the bioavailability of the chemical and increase its excretion rate. Chirality often plays an important role such as in the case of (R)-2-ethylhexanoic acid, which is a known teratogenic substance, while it's (S)-2-ethylhexanoic acid counterpart is not.[40b] Labile functional groups are to be avoided as they often dissociate to hazard substrates. One example is the toxicity of aliphatic nitriles caused by the metabolic release of cyanide.[41] On the other hand, the introduction of electron-withdrawing functional groups to aromatic rings can decrease the potential toxicity of chemicals such as the case with primaquine derivatives.[40a] Often, blocking the metabolic site of a molecule with fluorine or some other functional group (chlorine, nitrile, bulky aliphatic group) will prevent unwanted activation. This is the case with 7-methyl-1-fluorobenz[α]anthracene where the F atom prevents epoxidation of the molecule eliminating carcinogenicity of the fluorine-less analog 7-methyl-benz[α]anthracene.[42] Other valid strategies include the incorporation of labile groups into the cyclic structure, changing the size of rings, and increasing steric hindrance to prevent bioactivation.

## 7.17 How to Design to Reduce Carcinogenicity and Mutagenicity?

Carcinogenic and mutagenic potentials of chemicals are some of the major chronic toxicity factors that should be considered when designing safer chemicals. While it is quite difficult to predict all of the adverse effects the chemicals will exhibit, some design strategies can be used to reduce these potentials. Bioactivation of molecules to electrophiles should be avoided, which can be achieved by increasing the molecular weight of the molecule above 1000 Da, and by modifying the molecular shape in such a way to avoid binding to the sites at proteins and nucleic acids. Figure 7.19 shows some design suggestions for reduced carcinogenicity and mutagenicity.

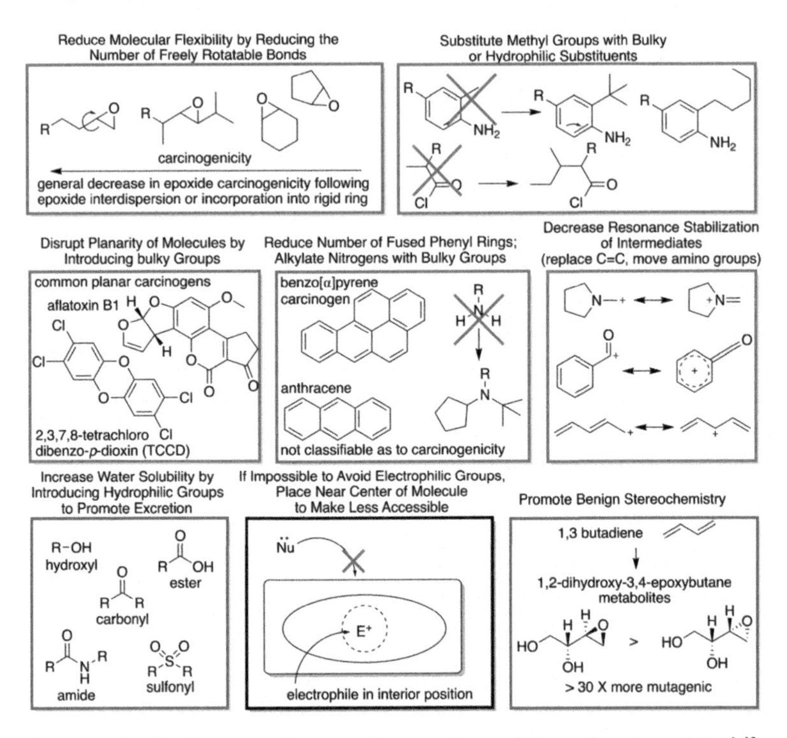


**Figure 7.19** Design strategies to reduce carcinogenicity and mutagenicity.[1,43]

Molecular flexibility, the feature often responsible for the bioavailability of the chemicals, can be limited by reducing the number of freely rotatable bonds. Epoxides generally exhibit decreased carcinogenicity when incorporated into more rigid cyclic structures. As the shape of the molecule has a great influence on the bioavailability of the molecule, smaller groups such as methyl can be replaced with bulkier or hydrophilic substituents. Due to the potential of planar molecules such as aflatoxin B1 to intercalate into DNA or RNA, the planarity of the molecule should be disrupted by introducing bulky groups. As we mentioned in previous chapters PAHs are known hazardous substances, thus one of the strategies to reduce their carcinogenicity is to reduce the number of fused phenyl rings. Bioactivation of a large number of molecules include resonance stabilization of intermediates that then exhibit carcinogenicity or mutagenicity; thus, this stabilization should be decreased by replacing –C=C– and $-CH_2-$ moieties with longer alkyl chains or carbonyl groups. Usually, when designing novel molecules, electrophilic functional groups should be avoided, but often this is not possible. In such cases, electrophilic groups should be placed close to the center of the molecule which will reduce molecular flexibility and reduce the accessibility of the electrophilic group to the cellular nucleophiles such as DNA. Additionally, benign stereochemistry should be promoted whenever possible. Meng et al. have shown that different stereochemical configurations of 1,3-butadiene epoxy metabolites exhibit different levels of mutagenicity.[43a]

One well-known example that shows how slight changes in molecular structure can cause a drastic change in the behavior of the chemical is the case of thalidomide.[1] This drug was introduced in 1956 as a morning sickness remedy for pregnant women, however, it was soon discovered that it caused teratogenic and mutagenic effects in newborns. It was shown that the S-enantiomer of thalidomide intercalates into DNA which causes interference with the mechanism of gene expression for several pathways. On the other hand, the R-enantiomer cannot bind with the DNA grove due to steric hindrance and acts as a sedative in the body, but it still exhibits teratogenic effect in rabbit models, albeit less than S-enantiomer. More recently, cereblon (CRBN) protein was accepted as a primary direct target of thalidomide binding.[44]

## 7.18 Specific Classes of Chemicals and Design Suggestions: Electrophiles

Chemicals with electrophilic character are generally not desirable in biological systems due to their tendency to irreversibly bind to macromolecules in the body. Thus, when considering design strategies towards safer chemicals, the molecules possessing electrophilic characteristics should be avoided. This can be done by avoiding using electrophilic groups or those functional groups that get metabolized to electrophiles, e.g., carbonium—benzyl, allyl, aryl groups, tert-alkyl halides, $\alpha, \beta$-unsaturated carbonyls, nitrenium and aziridium ions, epoxides and oxonium ions, aldehydes, peroxides, free radicals, quinones and quinone intermediates, acylating intermediates. If it is not possible to avoid the use of the electrophilic groups, it is possible to either introduce (bulky) substituents next to electrophilic groups or position the electrophilic group away from terminal positions and close to the middle of the molecule.

For the design strategies that can be used to modify different classes of chemicals, there are more specific guidelines that can be followed.

### 7.18.1 Aromatic Amines and Azo Dyes

Azo dyes are a large class of chemicals that are often used in the cosmetic, food, textile and leather, and pharmaceutical industries (Figure 7.20).[45] They are characterized by one or more azo bonds (R–N=N–R′) and often contain aromatic groups since their essential precursors are aromatic amines. The aromatic group conjugated with the double azo bond is the moiety that is mostly responsible for the color of the dye. This feature is also responsible for the concern surrounding the azo dyes since cleavage of the azo bond can result in the release of carcinogenic aromatic amines. For example, many azo dyes used in the pharmaceutical industry have shown toxic effects by causing allergic reactions, carcinogenicity or mutagenicity.[46] Their toxicity is directly related to the molecular structure and the degradation products.

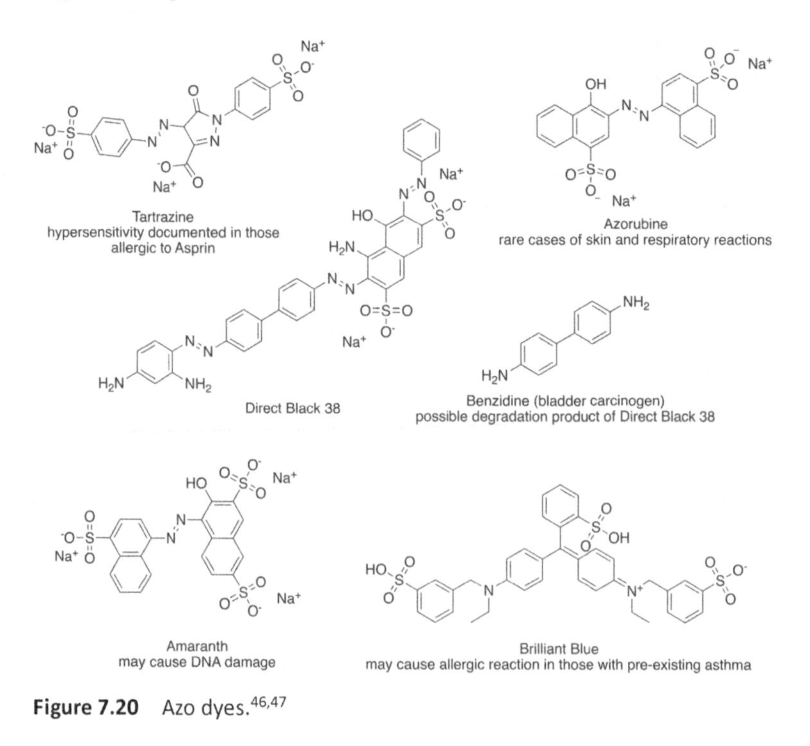


**Figure 7.20** Azo dyes.[46,47]

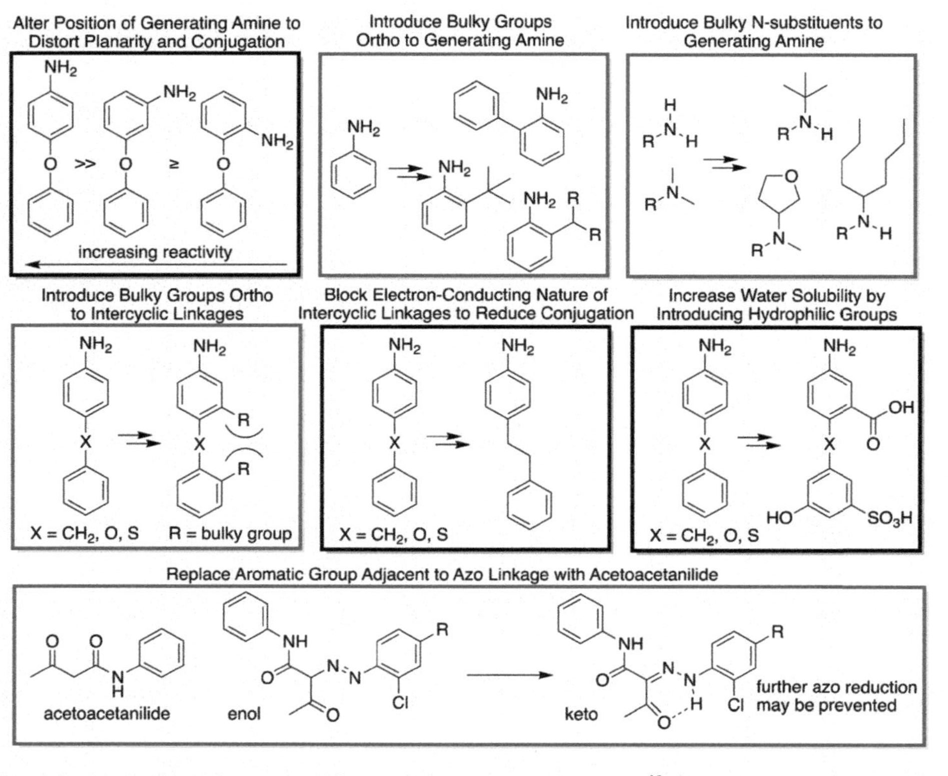


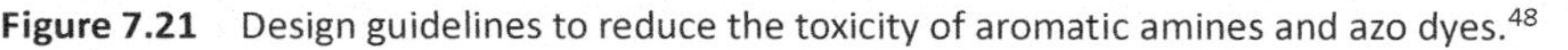

**Figure 7.21** Design guidelines to reduce the toxicity of aromatic amines and azo dyes.[48]

The structure of aromatic amines and azo dyes can be altered by following several design suggestions (Fig. 7.21). Bulky substituents should be introduced in the ortho position relative to the amine (or amine generating) groups which will provide steric hindrance and inhibit bioactivation. This can also be achieved by introducing bulky *N*-substituents to the amine. Bioavailability of the molecule can be reduced by distorting planarity which can be achieved by introducing bulky groups ortho to intercyclic linkages, and by positioning amine in the aromatic ring in a way that will reduce the stability of the electrophilic nitrenium ion. The conjugation can also be reduced by changing the electron-conducting nature of intercyclic linkages. Additionally, the release of aromatic amines from azo dyes can be prevented by replacing the aromatic group adjacent to the azo linkage with some other group that will hinder the release of the aromatic amine. One example is shown in Fig. 7.21 where the acetoacetanilide group is part of 3,3′-dichlorobenzidine molecule; the azo linkage is most likely removed by keto-enol tautomerization and thus, the azo reduction is prevented.[48]

## 7.18.2 Aldehydes and Substituted Aldehydes

Aldehydes are a class of molecules that are generally very reactive. As such, they usually exhibit toxic effects in the body ranging from cytotoxic, carcinogenic, genotoxic, to mutagenic effects.[49] Due to the structural diversity of aldehyde derivatives the mechanism of aldehyde toxicity is not very well understood.[50] However, it is possible to classify aldehydes based on their electronic characteristics, i.e., electrophilicity, which can determine their mechanism of toxicity, with the exception of some aldehydes where the solubility or steric hindrance can limit their target accessibility. As strong electrophiles, aldehydes can form toxic adducts in cellular proteins through addition reactions with amines and thiols. One of these, commonly associated with liver toxicity is depletion of glutathione when present in α,β-unsaturated configuration, which leads to covalent binding with cellular proteins.

To reduce the toxicity of aldehydes, the size of the alkyl chain can be increased to reduce crosslinking. Additionally, aldehyde group placement in a position to an aromatic system or double bond should be avoided so that the reactive group can be stabilized.

Design guidelines and some examples are shown in Figs. 7.22 and 7.23.

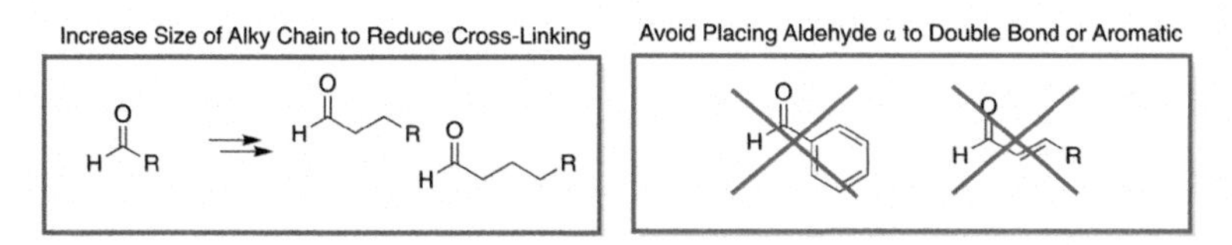


**Figure 7.22** Design guidelines to reduce the toxicity of aldehydes.

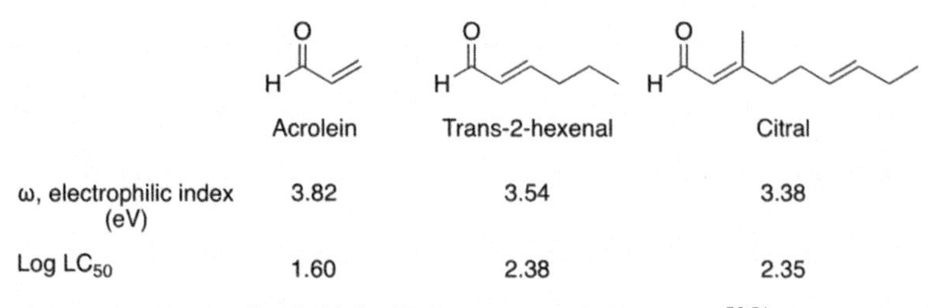

Acrolein
Trans-2-hexenal
Citral
ω, electrophilic index (eV)
3.82
3.54
3.38
Log $LC_{50}$
1.60
2.38
2.35


**Figure 7.23** Example of aldehyde changes to reduce toxicity.[50,51]

## 7.18.3 Acylating Agents and Isocyanates

The acylating agents are compounds known as carcinogens due to the formation of reactive acylium ions intermediates (Fig. 7.24). If the carbonyl group has electron-withdrawing groups on it, the positive charge on the carbon will be enhanced. Dimethylcarbamyl chloride (DMCC) hydrolyzes rapidly forming reactive acylium ion that reacts readily with the nucleic bases of the DNA.[52] On the other hand, diethylcarbamyl chloride was shown to be a less potent carcinogen, while dichloroacetyl chloride and ethylchloroformate are much weaker carcinogens compared to the DMCC due to decreased stability influenced by electronic effects. Another potent and reactive group of chemicals are isocyanates since the isocyanate group can form urethane linkage when reacting with the hydroxyl functional group. The resulting molecules are easily hydrolyzed and short-lived.

As these compounds are potent acylating agents due to the electrophilic property of the acyl group, increasing the steric hindrance adjacent to the carbamoyl moiety and avoiding the inclusion of electron-withdrawing groups on the carbonyl moiety will reduce their acylating potential (see Fig. 7.25).

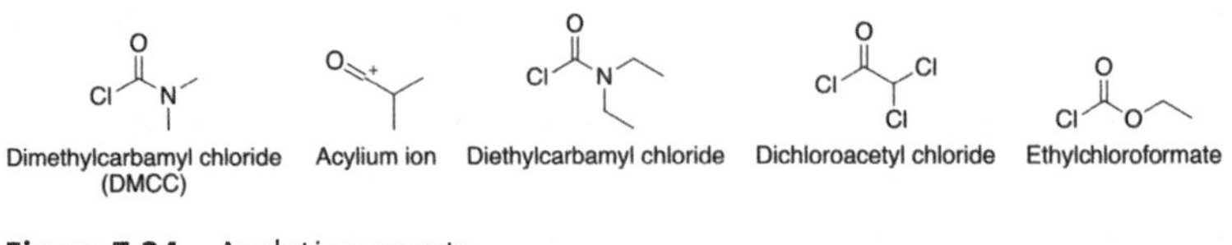

**Figure 7.24** Acylating agents.

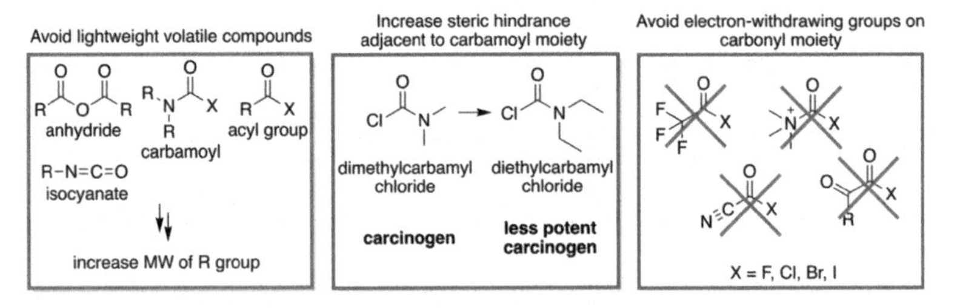


**Figure 7.25** Design guidelines to reduce the toxicity of acylating agents and isocyanates.

## 7.18.4 Alkyl Esters of Strong Acids

Alkylating activity of alkyl esters of strong acids such as sulfate, phosphate, or methanesulfonate depends on the size of the alkyl chain (Fig. 7.26). The reactivity is decreasing with the size and greatly diminishes beyond the butyl group.

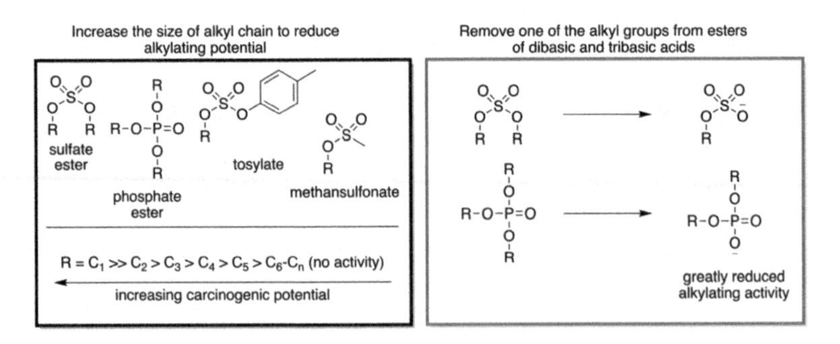


**Figure 7.26** Design guidelines to reduce the toxicity of alkyl esters of strong acids.[48]

Additionally, in the case of alkyl esters of dibasic (sulfate) and tribasic (phosphate) acids, if one of the alkyl groups is removed (hydrolyzed) the alkylating activity is greatly diminished and can be even completely removed (Fig. 7.26).[48]

### 7.18.5 Aliphatic Azo, Azoxy, and Hydrazo Compounds

Hydrazo compounds are metabolically activated when forming carcinogenic azoxyalkanes and azoalkanes. Most of these compounds that were tested for carcinogenic activity were found as highly carcinogenic.[53] However, it is possible to reduce the toxicity of these compounds by replacing alkyl groups with either bulky or polar substituents, as this will disrupt their metabolic activation (see Fig. 7.27).

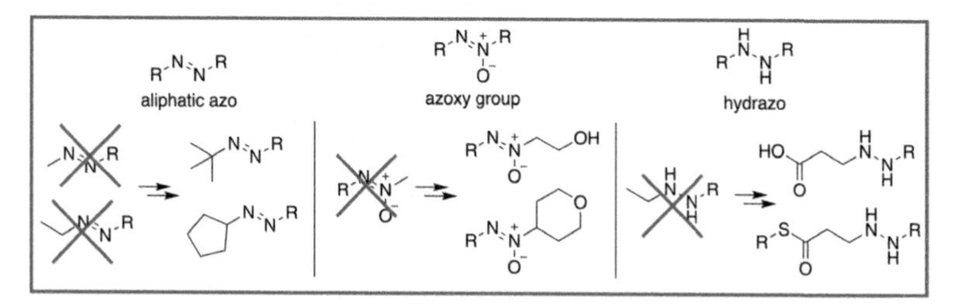


**Figure 7.27** Design guidelines to reduce the toxicity of aliphatic azo, azoxy, and hydrazo compounds.

### 7.18.6 Carbamates

The carbamates are esters of carbamic acids, usually used as insecticides. Similar to organophosphates, carbamates inhibit acetylcholinesterase (AChE) at nerve synapses by reversible binding. Toxic exposure can occur via ingestion, inhalation, or dermal exposure. Ethyl and vinyl carbamates are potent carcinogens.[54]

To counter the toxicity of carbamates, alkylating activity can be limited by modifying potential electrophilic sites in its structure. The alkyl groups at the carboxyl end can be replaced with longer alkyl chains as illustrated in Fig. 7.28. Ethyl carbamate is known to exhibit a teratogenic effect while butyl carbamate is non-teratogenic in hamsters.[55] Additionally, the alkoxy group should not be replaced by halogens (except F) at terminal or vicinal positions due to the cross-linking potential. Replacement of the alkoxy group in ethyl carbamate with chlorine yields dimethylcarbamyl chloride, a potent carcinogen. An increase of the steric (and electronic) hindrance of alkyl groups on the amino end will reduce the toxicity, since the formation of a good leaving group, particularly if it is an electrophile or electron-withdrawing group, can generate electrophilic intermediate and enhance alkylating activity.

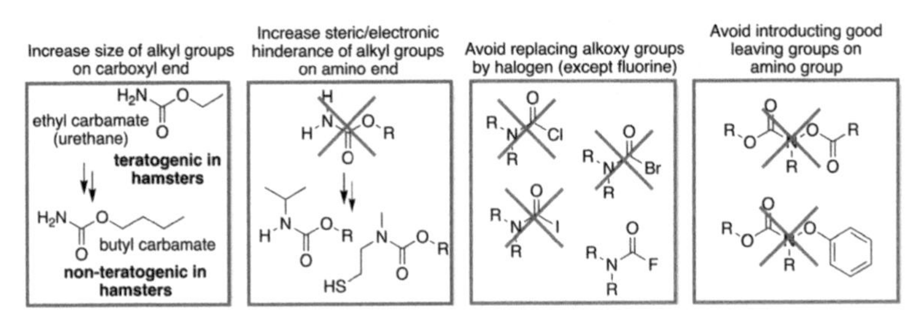


**Figure 7.28** Design guidelines to reduce the toxicity of carbamates.[55]

## 7.18.7 Epoxides and Ethylenimines

As mentioned in Chapter 6, epoxides are very reactive molecules possessing highly strained cyclic ether moiety consisting of a three-atom ring. Due to their electrophilic and lipophilic nature, they can cause adverse effects by interacting with nucleophiles such as amino acids and nucleic bases containing $OH^-$, $S^-$, $Cl^-$, or $NH_2$ groups.

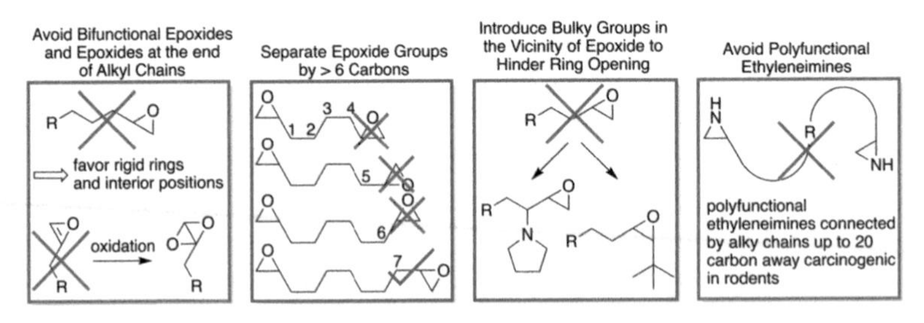


**Figure 7.29** Design guidelines to reduce the toxicity of epoxides and ethylenimines.[48]

If epoxide or ethylenimine moiety is necessary to be included in the molecular structure, several design suggestions can be utilized to reduce the toxicity of the designed chemical (Fig. 7.29). Small rings of epoxides and ethylenimine tend to open easily and generate electrophilic intermediates. Thus, when designing with epoxides in mind, the terminal position of the ring on the alkyl chain should be avoided and instead, epoxide should be attached to a more rigid ring (Fig. 7.19). Another way to prevent the opening of the ring is to include bulky functional groups in the vicinity of the epoxide. Often epoxides contain double bonds or can be bifunctional which can result in the formation of additional reactive epoxide; thus, it should be avoided. When the molecule contains two epoxide groups, they should be separated by at least six carbon atoms to decrease carcinogenicity. If the ethylenimines are included in the molecular structure, they should be kept monofunctional, as polyfunctional ethyleneimines are known to be more carcinogenic.

## 7.18.8 Lactones

Lactones can easily form reactive electrophilic intermediates by opening rings and generating carbonium ions and acylating intermediates. Valid design strategies for reduced toxicity of lactones revolve around the prevention of the ring-opening (Fig. 7.30). Thus, if the rings are larger than six-membered rings, ring strain will be reduced and can resist opening. Another way to stabilize the ring is to add bulky or hydrophilic substituents to the ring, which will also decrease the absorption of the molecule. On the other hand, if the double bond is part of the ring, especially in the vicinity of the carbonyl group, the activity can be restored through resonance stabilization of carbonium ion and increased crosslinking potential.

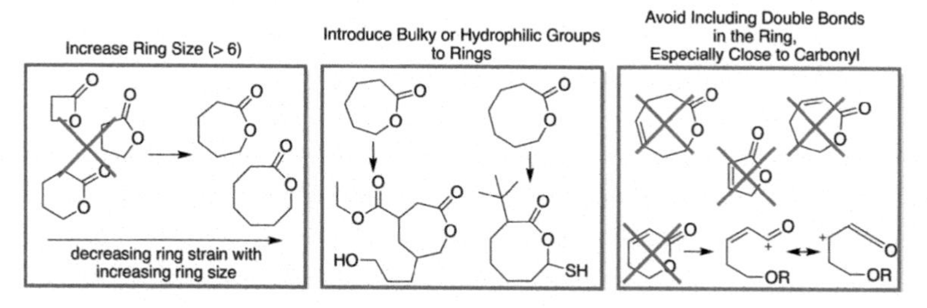


**Figure 7.30** Design guidelines to reduce the toxicity of lactones.[48]

### 7.18.9 Haloalkanes and Substituted Haloalkanes

Since halogens are good leaving groups, haloalkanes with one halogen (except fluorine), can potentially act as alkylating agents. Both haloalkanes and haloalkenes can be involved in the direct and indirect mechanism of activation leading to carcinogenic activity. The alkylating potential of the haloalkanes decreases with the size of the halogen at the terminal carbon.

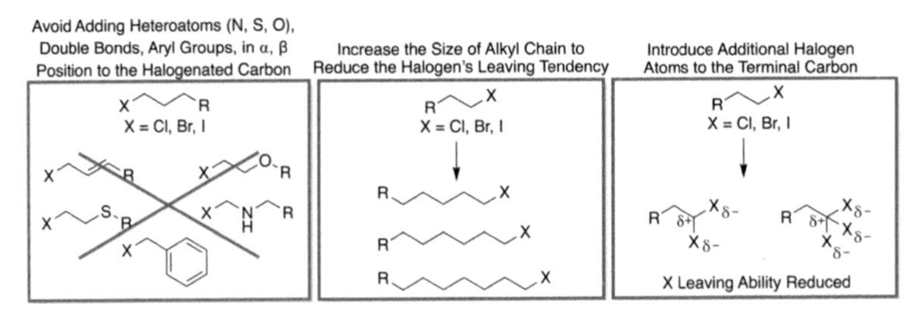


**Figure 7.31** Design guidelines to reduce the toxicity of haloalkanes.[48]

When designing safer haloalkanes, some structural precautions can be taken to reduce toxicity (see Fig. 7.31). Inclusion of heteroatoms (e.g., N, O, S), double bonds, or aryl groups in α- or β-positions relative to carbon with halogen substituent can increase the alkylating activity of haloalkanes. This occurs because the additional electrophilic group is introduced, or the existing halogen atom can leave easier. Leaving the halogen can be impeded by increasing the size of the alkyl chain, adding additional halogen atoms to terminal carbon, or by replacing halogens with others by considering their leaving tendency (I > Br >> Cl).

### 7.18.10 Michael Addition Acceptors

Molecules containing carbonyl, phosphoryl or sulfonyl group at the α-carbon of the terminal vinyl group are potential hazards since they can act as electrophiles. This is due to the possibility that a double bond can be polarized which causes a partial positive charge on the terminal carbon. This is particularly concerning when more than one Michael acceptor is present due to potential crosslinking. To counter this, bifunctional or multifunctional molecules containing more than one Michael acceptor should generally be avoided. Additionally, since the substitution on the vinyl group strongly influences the ability of the group to act as Michael acceptor, alkyl or other bulky substituents can be introduced to impair this ability via electronic or steric hindrance effect. These design suggestions are illustrated in Fig. 7.32.

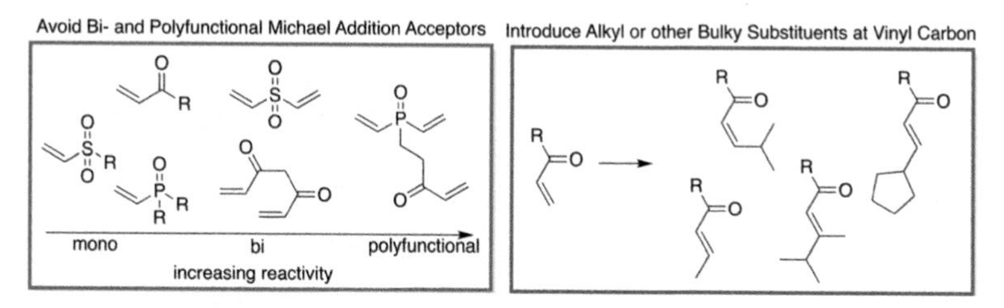


**Figure 7.32** Design guidelines to reduce the toxicity of Michael addition acceptors.[48]

## 7.18.11 *N*-nitrosamines

The predominant metabolic activation pathway for *N*-nitrosamines is α-hydroxylation, and most of these compounds that were tested for carcinogenicity were found to be carcinogenic.[56] Thus, the α-hydrogen is directly responsible for the *N*-nitrosamines activation. However, if the vicinity of α-carbon is sterically or electronically hindered, carcinogenic activity is significantly reduced. From the design perspective, this hindrance can be introduced by including branched alkyl or other bulky groups in the α-carbon vicinity (see Fig. 7.33). Another strategy is to include acidic, fluoro, or other bulky substituents on the α-carbon which will impair metabolic activation.

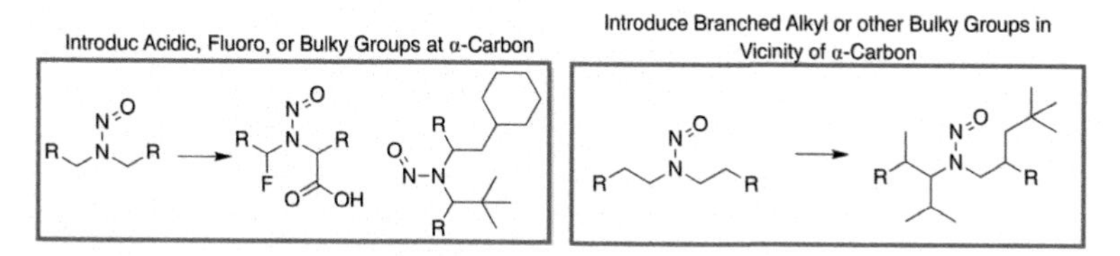


**Figure 7.33** Design guidelines to reduce the toxicity of *N*-nitrosamines.[48]

### 7.18.12 Organic Peroxides

Homolytic cleavage of the O-O bond results in the formation of free radicals such as alkoxy ($RO^•$), acyl ($RCO_2^•$), hydroxyl ($^•OH$), and finally alkyl ($R^•$) radical. Acting as strong oxidizing agents, organic peroxides can act as potent mutagenic and carcinogenic agents. This is most likely due to the high stability of the formed radical species that are also highly reactive. The carcinogenicity and genotoxicity of organic peroxides[57] revealed that small alkyl hydroperoxides and *tert*-dialkyl peroxides have much stronger mutagenic and carcinogenic potential compared to the other peroxides. Thus, when designing safer organic peroxides (as shown in Fig. 7.34), bulkier alkyl groups and longer alkyl chains can be introduced in alkyl hydroperoxides or dialkyl peroxides, while *tert*-butyl group should be avoided.

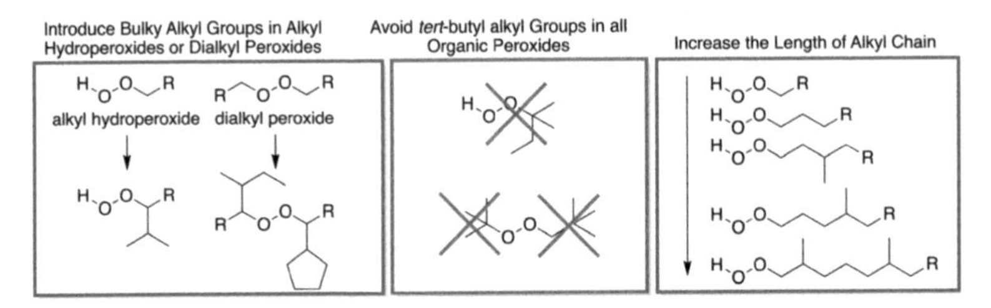


**Figure 7.34** Design guidelines to reduce the toxicity of organic peroxides.[48]

### 7.18.13 Organophosphorus Compounds

Organophosphorus compounds are widely used as flame retardants, pesticides, plasticizers, etc. However, they are known to cause neurotoxicity that is lethal in insects, nematodes and mammals. The lethality stems from the inhibition of acetylcholinesterase (AChE) (see Fig. 7.35). Most of the tested methyl or ethyl esters of phosphoric and thiophosphoric acids are highly electrophilic and exhibit mutagenic effects.[48]

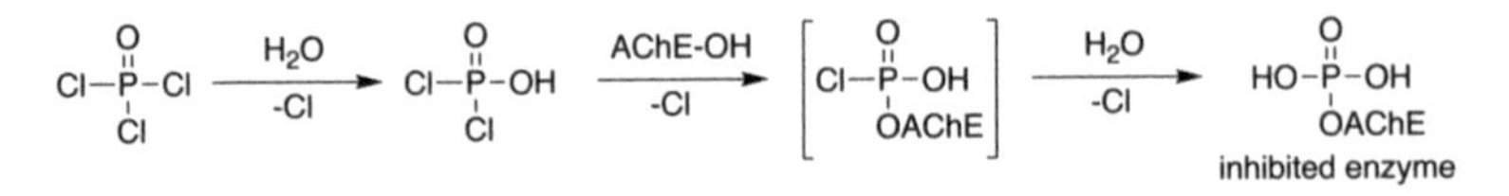


**Figure 7.35** Phosphorylation of AChE by $POCl_3$.[58]

When designing safer organophosphorus compounds (see Fig. 7.36), halogenation of vicinal or terminal positions should be avoided as this can lead to crosslinking and increased electrophilic reactivity of the breakdown products. The alkylating reactivity of organophosphorus compounds decreases with the size of the alkyl group, thus steric hindrance and hydrophilicity of the alkyl substituents should be promoted. On the other hand, the addition of electrophilic (electron-withdrawing) groups increases the alkylating reactivity and should be avoided. These points are illustrated in Fig. 7.36.

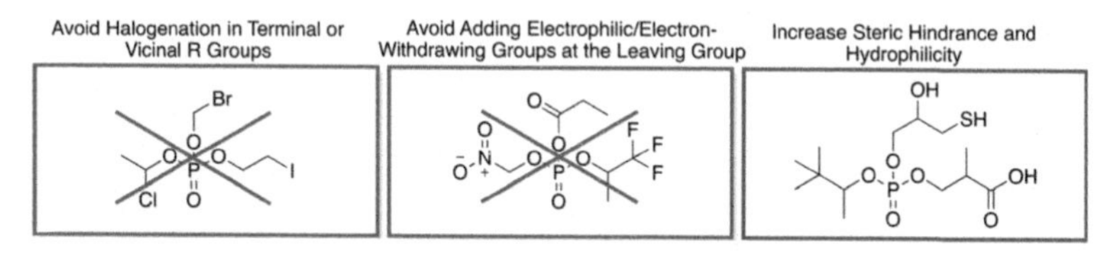
Avoid Halogenation in Terminal or Vicinal R Groups
Avoid Adding Electrophilic/Electron-Withdrawing Groups at the Leaving Group
Increase Steric Hindrance and Hydrophilicity

**Figure 7.36** Design guidelines to reduce the toxicity of organophosphorus compounds.[48]

## 7.18.14 Polycyclic Aromatic Hydrocarbons

Polycyclic aromatic hydrocarbons can cause carcinogenicity by intercalating into DNA, or by activation by CYP enzyme to dihydrodiol peroxides. The carcinogenicity can be affected by the number of fused phenyl rings, methyl substitutions, and steric shape/volume of the molecule. Especially important is the bay/fjord region, which can stabilize reactive epoxides giving it time to reach DNA and bind to it. The toxicity of several PAHs is shown in Fig. 7.37.

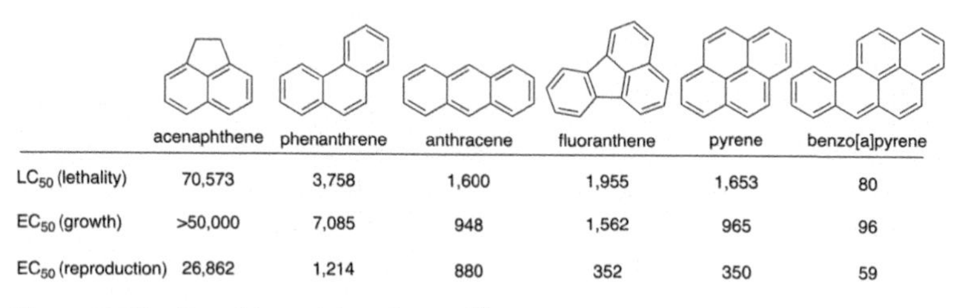

| | acenaphthene | phenanthrene | anthracene | fluoranthene | pyrene | benzo[a]pyrene |
|---|---|---|---|---|---|---|
| $LC_{50}$ (lethality) | 70,573 | 3,758 | 1,600 | 1,955 | 1,653 | 80 |
| $EC_{50}$ (growth) | >50,000 | 7,085 | 948 | 1,562 | 965 | 96 |
| $EC_{50}$ (reproduction) | 26,862 | 1,214 | 880 | 352 | 350 | 59 |

**Figure 7.37** Trend in toxicity of PAHs.[59]

The shape and size of the PAHs are crucial for their carcinogenicity. Most of the PAH carcinogens have 4–6 rings arranged in an angled manner forming a bay region. Their reactivity stems from the formation of bay diol epoxides in metabolism which, upon opening of the epoxide ring, generate electrophilic carbonium ions stabilized by the aromatic system. To counter this, the bay region can be blocked by ring fusion or the addition of a methyl group in the vicinity of the bay region. Other ways to reduce the toxicity of PAHs include introducing linear shape, where more than four rings will be fused, or limiting the number of fused rings to less than four; generating a high degree of symmetry; introducing acidic or bulky groups to the rings; and considering isosteric replacement of substituents with fluorine (as illustrated in Fig. 7.38).

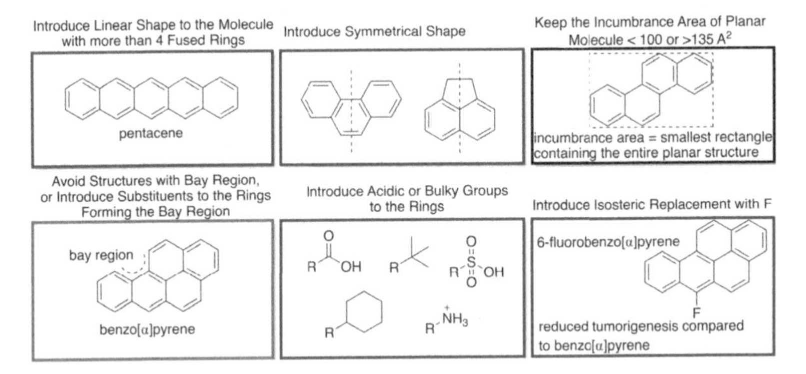


**Figure 7.38** Design guidelines to reduce the toxicity of polycyclic aromatic hydrocarbons.[59,60]

## 7.18.15 Per- and Polyfluoroalkyl Substances

Per- and polyfluoroalkyl substances have widely been used due to their characteristic of extreme stability of the C–F bond. However, as we mentioned in Chapter 3, they are very persistent and can bioaccumulate/bioconcentrate and have been linked with kidney cancer and thyroid disease. PFOA and PFAS were found to suppress the body's response to vaccines.[61] To reduce the endocrine disruption, the PFAs carbon chain length should be kept either short with less than 6 C atoms, or long with more than 10 C atoms. In the case of sulfonates, the carbon chain should be longer with more than

8 C atoms. Additionally, the number of F atoms should be reduced, by omitting them or replacing them with OH groups. It was noticed that perfluorobutanesulfonate (PFBS) and its related alcohols have less tendency to bioaccumulate and are less genotoxic compared to PFOA and PFOS.[62] These design suggestions are illustrated in Fig. 7.39.

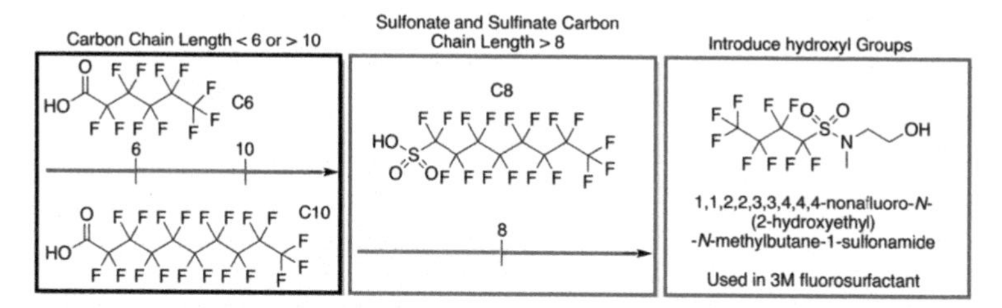


**Figure 7.39** Design guidelines to reduce the toxicity of polyfluoroalkyl substances.[63]

## 7.18.16 Quinones and Quinone-like Compounds

Quinones, such as benzoquinone and hydrobenzoquinone, conjugate to GSH to form mono, bis, tris, or tetra GS conjugates (Fig. 7.40).[64] Bis and tris forms are potent toxic agents, especially in the kidneys. They are highly electrophilic and cause oxidative stress for the cell through redox cycling which contributes to both homeostasis and cytotoxicity. Partially substituted quinones additionally undergo Michael adduct formation that can form covalent bonds with nucleophiles such as sulfur-containing amino acids.[65]

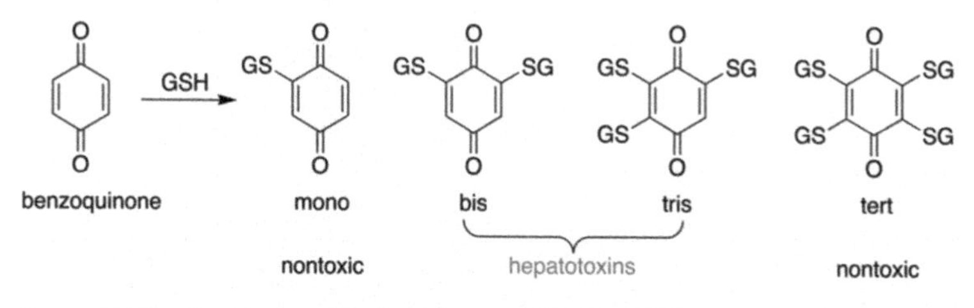


**Figure 7.40** Conjugate of hydroquinone with GSH.[65,66]

Since quinones are highly reactive, they should be avoided as much as possible, however, some strategies can be used to decrease their reactivity, and by that, toxicity (see Fig. 7.41). Due to the additional stabilization of free radical species upon oxidation of ortho/para hydroquinones (or quinone imines), the carbonyl/imine group should be moved to meta position from ortho/para position. Additionally, the introduction of phenyl rings or larger aromatic systems on quinones can lead to reduced toxicity (see Fig. 7.41).

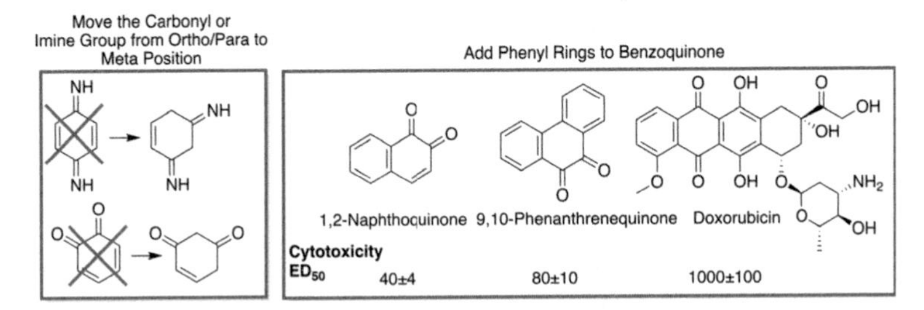


**Figure 7.41** Design guidelines to reduce the toxicity of quinones.[43b,64]

## References

1. Voutchkova, A. M.; Osimitz, T. G.; Anastas, P. T. Toward a comprehensive molecular design framework for reduced hazard. *Chem Rev*, 2010, **110**(10), 5845–5882.
2. (a) Voutchkova, A. M.; Kostal, J.; Steinfeld, J. B.; Emerson, J. W.; et al. Towards rational molecular design: derivation of property guidelines for reduced acute aquatic toxicity. *Green Chem*, 2011, **13**(9), 2373–2379; (b) Kostal, J.; Voutchkova-Kostal, A.; Anastas, P. T.; Zimmerman, J. B. Identifying and designing chemicals with minimal acute aquatic toxicity. *Proc Natl Acad Sci*, 2015, **112**(20), 6289–6294.
3. Newsome, L. D.; Nabholz, J. V.; Kim, A. Designing aquatically safer chemicals. In: Stephen C, D.; Garrett, R. L. (Eds.), *Designing Safer Chemicals*, Vol. 640, American Chemical Society: Washington, DC, 1996, pp. 172–192.
4. (a) Ma, T. *Ignitability and Explosibility of Gases and Vapors*. Springer: New York, 2015; (b) Mumford, C. H.; Carson, P. A. *Hazardous Chemicals Handbook*, 2nd ed.; Butterworth-Heinemann: 2002, p. 640.
5. de Wit, C. A. An overview of brominated flame retardants in the environment. *Chemosphere*, 2002, **46**(5), 583–624.
6. Meerts, I. A.; Letcher, R. J.; Hoving, S.; Marsh, G.; et al. In vitro estrogenicity of polybrominated diphenyl ethers, hydroxylated PDBEs,

and polybrominated bisphenol A compounds. *Environ Health Perspect*, 2001, **109**(4), 399–407.

7. Johnson-Restrepo, B.; Kannan, K. An assessment of sources and pathways of human exposure to polybrominated diphenyl ethers in the United States. *Chemosphere*, 2009, **76**(4), 542–548.
8. United States Environmental Protection Agency. An alternative assessment for the flame retardant decabromodipenyl ether (DecaBDE). U. S. EPA: 2014, p. 901.
9. (a) Kolibaba, T. J.; Shih, C.-C.; Lazar, S.; Tai, B. L.; et al. Self-extinguishing additive manufacturing filament from a unique combination of polylactic acid and a polyelectrolyte complex. *ACS Mater Lett*, 2019, **2**(1), 15–19; (b) Leistner, M.; Haile, M.; Rohmer, S.; Abu-Odeh, A.; et al. Water-soluble polyelectrolyte complex nanocoating for flame retardant nylon-cotton fabric. *Polym Degrad Stab*, 2015, **122**, 1–7; (c) Haile, M.; Fincher, C.; Fomete, S.; Grunlan, J. C. Water-soluble polyelectrolyte complexes that extinguish fire on cotton fabric when deposited as pH-cured nanocoating. *Polym Degrad Stab*, 2015, **114**, 60–64.
10. Malucelli, G. Biomacromolecules and bio-sourced products for the design of flame retarded fabrics: current state of the art and future perspectives. *Molecules*, 2019, **24**(20), 3774.
11. Oxley, J. C. The chemistry of explosives. In: Zukas, J. A.; Walters, W. P. (Eds.), *Explosive Effects and Applications*, Springer: New York, NY, 1998, pp. 137–172.
12. Leo, A. J.; Hoekman, D. Calculating log P(oct) with no missing fragments; The problem of estimating new interaction parameters. *Perspect drug discov des*, 2000, **18**(1), 19–38.
13. (a) Abraham, M. H.; Chadha, H. S.; Leitao, R. A. E.; Mitchell, R. C.; et al. Determination of solute lipophilicity, as log P(octanol) and log P(alkane) using poly(styrene–divinylbenzene) and immobilised artificial membrane stationary phases in reversed-phase high-performance liquid chromatography. *J Chromatogr A*, 1997, **766**(1–2), 35–47; (b) Abraham, M. H.; Gola, J. M. R.; Kumarsingh, R.; Cometto-Muniz, J. E.; et al. Connection between chromatographic data and biological data. *J Chromatogr B Biomed Sci Appl*, 2000, **745**(1), 103–115.
14. (a) Information, N. C. f. B. PubChem compound summary for CID 6568, 2-butanol. https://pubchem.ncbi.nlm.nih.gov/compound/2-Butanol (accessed June 28); (b) Information, N. C. f. B. PubChem compound summary for CID 263, 1-butanol. https://pubchem.ncbi.nlm.nih.

gov/compound/1-Butanol (accessed June 28); (c) Information, N. C. f. B. PubChem compound summary for CID 887, methanol. https://pubchem.ncbi.nlm.nih.gov/compound/Methanol (accessed June 28); (d) Information, N. C. f. B. PubChem compound summary for CID 962, water. https://pubchem.ncbi.nlm.nih.gov/compound/Water (accessed June 28); (e) Information, N. C. f. B. PubChem compound summary for CID 402, hydrogen sulfide. https://pubchem.ncbi.nlm.nih.gov/compound/Hydrogen-sulfide (accessed June 28); (f) Information, N. C. f. B. PubChem compound summary for CID 8064, 1,4-butanediol. https://pubchem.ncbi.nlm.nih.gov/compound/1_4-Butanediol (accessed June 28); (g) Information, N. C. f. B. PubChem compound summary for CID 8007, butylamine. https://pubchem.ncbi.nlm.nih.gov/compound/Butylamine (accessed June 28).

15. Labiris, N. R.; Dolovich, M. B. Pulmonary drug delivery. Part I: physiological factors affecting therapeutic effectiveness of aerosolized medications. *Br J Clin Pharmacol*, 2003, **56**(6), 588–599.
16. Lehman-McKeeman, L. D. Absorption, distribution, and excretion of toxicants. In: Klaassen, C. D.; Watkins, III, J. B. (Eds.), *Casarett & Doull's Essentials of Toxicology*, 3rd ed., McGraw-Hill Education: New York, 2015, pp. 61–77.
17. Xiang, T. X.; Anderson, B. D. Influence of a transmembrane protein on the permeability of small molecules across lipid membranes. *J Membr Biol*, 2000, **173**(3), 187–201.
18. (a) Kontoghiorghes, G. J. Chelators affecting iron absorption in mice. *Arzneimittelforschung*, 1990, **40**(12), 1332–1335; (b) Trendeleva, T. A.; Sukhanova, E. I.; Rogov, A. G.; Zvyagilskaya, R. A.; et al. Role of charge screening and delocalization for lipophilic cation permeability of model and mitochondrial membranes. *Mitochondrion*, 2013, **13**(5), 500–506.
19. Sugano, K.; Nabuchi, Y.; Machida, M.; Asoh, Y. Permeation characteristics of a hydrophilic basic compound across a bio-mimetic artificial membrane. *Int J Pharm*, 2004, **275**(1), 271–278.
20. N'Da, D. D. Prodrug strategies for enhancing the percutaneous absorption of drugs. *Molecules*, 2014, **19**(12), 20780–20807.
21. Billich, A.; Vyplel, H.; Grassberger, M.; Schmook, F. P.; et al. Novel cyclosporin derivatives featuring enhanced skin penetration despite increased molecular weight. *Bioorg Med Chem*, 2005, **13**(9), 3157–3167.
22. Pugh, W. J.; Roberts, M. S.; Hadgraft, J. Epidermal permeability—penetrant structure relationships: 3. The effect of hydrogen bonding

interactions and molecular size on diffusion across the stratum corneum. *Int J Pharm*, 1996, **138**(2), 149–165.

23. Jafari, F.; Moradi, S.; Nowroozi, A.; Sadrjavadi, K.; et al. Exploring the binding mechanism of paraquat to DNA by a combination of spectroscopic, cellular uptake, molecular docking and molecular dynamics simulation methods. *N J Chem*, 2017, **41**(23), 14188–14198.
24. Watkinson, R. J.; Morgan, P. Physiology of aliphatic hydrocarbon-degrading microorganisms. *Biodegradation*, 1990, **1**(2–3), 79–92.
25. Lloyd-Jones, G.; Trudgill, P. W. The degradation of alicyclic hydrocarbons by a microbial consortium. *Int Biodeterior*, 1989, **25**(1–3), 197–206.
26. Boethling, R. S.; Sommer, E.; DiFiore, D. Designing small molecules for biodegradability. *Chem Rev* 2007, **107**(6), 2207–2227.
27. Howard, P. H.; Boethling, R. S. Designing for non-persistence. In: Anastas, P. T.; Boethling, R.; Voutchkova, A. (Eds.) *Handbook of Green Chemistry: Designing Safer Chemicals*, Vol. 9, Wiley-VCH: Weinheim, Germany, 2012, pp. 453–484.
28. (a) Shiraki, S.; Garcia-Garibay, M. A. Carbon–carbon bond formation by the photoelimination of small molecules in solution and in crystals. In: Fagnoni, M.; Albini, A. (Eds.), *Handbook of Synthetic Photochemistry*, Wiley-VCH: Weinheim, Germany, 2010, pp. 25–66; (b) Passananti, M.; Lavorgna, M.; Iesce, M. R.; DellaGreca, M.; et al. Chlorpropham and phenisopham: phototransformation and ecotoxicity of carbamates in the aquatic environment. *Environ. Sci.: Process. Impacts*, 2014, **16**(4), 823–831; (c) Bochet, C. G.; Blanc, A. Photolabile protecting groups in organic synthesis. In: Fagnoni, M.; Albini, A. (Eds.), *Handbook of Synthetic Photochemistry*, Wiley-VCH: Weinheim, Germany, 2010, pp. 417–447.
29. Drossman, H.; Johnson, H.; Mill, T. Structure activity relationships for environmental processes 1: hydrolysis of esters and carbamates. *Chemosphere*, 1988, **17**(8), 1509–1530.
30. Beadham, I.; Gurbisz, M.; Gathergood, N. Design of safer chemicals: ionic liquids. In: Anastas, P. T.; Boethling, R.; Voutchkova, A. (Eds.), *Handbook of Green Chemistry: Designing Safer Chemicals*, Vol. 9, Wiley-VCH: Weinheim, Germany, 2012, pp. 137–158.
31. Brinck, T.; Haeberlein, M.; Jonsson, M. A computational analysis of substituent effects on the O–H bond dissociation energy in phenols: polar versus radical effects. *J Am Chem Soc*, 1997, **119**(18), 4239–4244.
32. (a) Alexander, D. E. Bioaccumulation, bioconcentration, biomagnification. In: Alexander, D. E.; Fairbridge, R. W. (Eds.),

*Environmental Geology*, Springer: Dordrecht, 1999, pp. 43–44; (b) Niki, E.; Abe, K. Vitamin E: structure, properties and functions. In: Niki, E. (Ed.), *Vitamin E: Chemistry and Nutritional Benefits*, The Royal Society of Chemistry: Croydon, UK, 2019, pp. 1–11; (c) Morel, F. M. M.; Kraepiel, A. M. L.; Amyot, M. The chemical cycle and bioaccumulation of mercury. *Ann Rev Ecol Syst*, 1998, **29**(1), 543–566; (d) Krabbenhoft, D. P.; Rickert, D. A. Mercury contamination of aquatic ecosystems. https://pubs.usgs.gov/fs/1995/fs216-95/ (accessed January 15); (e) Kodavanti, P. R. S.; Loganathan, B. G. Organohalogen pollutants and human health. In: Quah, S. R. (Ed.), *International Encyclopedia of Public Health*, 2nd ed., Academic Press: Oxford, 2017, pp. 359–366; (f) Neely, W. B.; Branson, D. R.; Blau, G. E. Partition coefficient to measure bioconcentration potential of organic chemicals in fish. *Environ Sci Tech*, 1974, **8**(13), 1113–1115.

33. Jones, J. P. Metabolic concerns in drug design. In: Elfarra, A. (Ed.), *Biotechnology—Pharmaceutical Aspects: Advances in Bioactivation Research*, Vol. 9, Springer: New York, NY, 2008, pp. 3–26.

34. (a) Walsh, J. S.; Miwa, G. T. Bioactivation of drugs: risk and drug design. *Ann Rev Pharmacol Toxicol*, 2011, **51**(1), 145–167; (b) Limban, C.; Nuţă, D. C.; Chiriţă, C.; Negreş, S.; et al. The use of structural alerts to avoid the toxicity of pharmaceuticals. *Toxicol Rep*, 2018, **5**, 943–953.

35. Von Moltke, L. L.; Greenblatt, D. J.; Granda, B. W.; Duan, S. X.; et al. Zolpidem metabolism *in vitro*: responsible cytochromes, chemical inhibitors, and *in vivo* correlations. *Br J Clin Pharmacol*, 1999, **48**(1), 89–97.

36. Tingle, M. D.; Jewell, H.; Maggs, J. L.; O'Neill, P. M.; et al. The bioactivation of amodiaquine by human polymorphonuclear leucocytes *in vitro*: chemical mechanisms and the effects of fluorine substitution. *Biochem Pharmacol*, 1995, **50**(7), 1113–1119.

37. Fu, P. P.; Xia, Q.; Lin, G.; Chou, M. W. Pyrrolizidine alkaloids—genotoxicity, metabolism enzymes, metabolic activation, and mechanisms. *Drug Metab Rev*, 2004, **36**(1), 1–55.

38. Kerns, E. H.; Di, L. *Drug-like Properties: Concepts, Structure Design and Methods*. Elsevier: Burlington, MA, 2008, p. 526.

39. (a) Lipinski, C. A.; Lombardo, F.; Dominy, B. W.; Feeney, P. J. Experimental and computational approaches to estimate solubility and permeability in drug discovery and development settings. *Adv Drug Deliver Rev*, 1997, **23**(1–3), 3–25; (b) Veber, D. F.; Johnson, S. R.; Cheng, H.-Y.; Smith, B. R.; et al. Molecular properties that influence the oral bioavailability of drug candidates. *J Med Chem* 2002, **45**(12), 2615–2623.

40. (a) Liu, H.; Ding, Y.; Walker, L. A.; Doerksen, R. J. Computational study on the effect of exocyclic substituents on the ionization potential of primaquine: insights into the design of primaquine-based antimalarial drugs with less methemoglobin generation. *Chem Res Toxicol*, 2015, **28**(2), 169–174; (b) DeVito, S. C. Structural and toxic mechanism-based approaches to designing safer chemicals. In: Anastas, P. T.; Boethling, R.; Voutchkova, A. (Eds.), *Handbook of Green Chemistry*, Wiley-VCH: Weinheim, Germany, 2012, pp. 77–106.
41. Willhite, C. C.; Smith, R. P. The role of cyanide liberation in the acute toxicity of aliphatic nitriles. *Toxicology and Applied Pharmacology* 1981, **59**(3), 589–602.
42. Ariëns, E. J. Design of safer chemicals. *Medicinal Chemistry*, 1980, **9**, 1–46. https://doi.org/10.1016/B978-0-12-060309-1.50007-0
43. (a) Meng, R. Q.; Hackfeld, L. C.; Hedge, R. P.; Wisse, L. A.; et al. Mutagenicity of stereochemical configurations of 1,3-butadiene epoxy metabolites in human cells. HEI Research Report 150. Health Effects Institute, Boston, MA; (b) Lai, D. Y.; Woo, Y.-t. Reducing carcinogenicity and mutagenicity through mechanism-based molecular design of chemicals. In: Anastas, P. T.; Boethling, R.; Voutchkova, A. (Eds.), *Handbook of Green Chemistry*, Vol. 9, Wiley-VCH: Weinheim, Germany, 2012, pp. 375–406.
44. Mori, T.; Ito, T.; Liu, S.; Ando, H.; et al. Structural basis of thalidomide enantiomer binding to cereblon. *Sci Rep*, 2018, **8**(1), 1294.
45. Hildenbrand, S.; Schmahl, F. W.; Wodarz, R.; Kimmel, R.; et al. Azo dyes and carcinogenic aromatic amines in cell cultures. *Internat Arch Occup Environ Health*, 1999, **72**(3), M052–M056.
46. Gičević, A.; Hindija, L.; Karačić, A. *Toxicity of azo dyes in pharmaceutical industry, IFMBE Proceedings*, Vol. 73, U.S. EPA, 2020, pp. 581–588.
47. Occupational Safety and Health Administration. Benzidine-based dyes: direct black 38, direct brown 95 and direct blue 6 dyes. OSHA Instruction, Act 5(a)(1), standard number 1910 1910.1003 1910.1010, 1980, United States Department of Labor.
48. Lai, D. Y.; Woo, Y.-T. Reducing carcinogenicity and mutagenicity through mechanism-based molecular design of chemicals. In: P. T. Anastas (Ed.), *Handbook of Green Chemistry: Designing Safer Chemicals*, Vol. 9, Wiley-VCH: Weinheim, Germany, 2012, pp. 375–406.
49. Laskar, A. A.; Younus, H. Aldehyde toxicity and metabolism: the role of aldehyde dehydrogenases in detoxification, drug resistance and carcinogenesis. *Drug Metab Rev*, 2019, **51**(1), 42–64.
50. LoPachin, R. M.; Gavin, T. Molecular mechanisms of aldehyde toxicity: a chemical perspective. *Chem Res Toxicol*, 2014, **27**(7), 1081–1091.

51. Chan, K.; Poon, R.; O'Brien, P. J. Application of structure–activity relationships to investigate the molecular mechanisms of hepatocyte toxicity and electrophilic reactivity of α,β-unsaturated aldehydes. *J Appl Toxicol*, 2008, **28**(8), 1027–1039.

52. Van Duuren, B. L. Direct-acting alkylating and acylating agents. *Ann N Y Acad Sci*, 1988, **534**(1), 620–634.

53. Arcos, J. C.; Woo, Y.-T.; Argus, M. F. *Aliphatic Carcinogens: Structural Bases and Biological Mechanisms*. Academic Press: New York, 1982, p. 780.

54. Silberman, J.; Taylor, A. *Carbamate Toxicity*. StatPearls: Treasure Island, FL, 2020.

55. DiPaolo, J. A.; Elis, J. The comparison of teratogenic and carcinogenic effects of some carbamate compounds. *Cancer Res*, 1967, **27**(9), 1696–1701.

56. (a) Woo, Y.; Lai, D. Y.; Arcos, J. C.; Argus, M. F. *Chemical Induction of Cancer: Structural Bases and Biological Mechanisms*, Vol. III B, *Aliphatic and Polyhalogenated Carcinogens*, 1985; (b) Miller, E. C. Some current perspectives on chemical carcinogenesis in humans and experimental animals: presidential address. *Cancer Res*, 1978, **38**, 1479–1496.

57. Lai, D. Y.; Woo, Y. t.; Argus, M. F.; Arcos, J. C. Carcinogenic potential of organic peroxides: prediction based on structure-activity relationships (SAR) and mechanism-based short-term tests. *Environ Carcinog Ecotoxicol Rev*, 1996, **14**(1), 63–80.

58. Quistad, G. B.; Zhang, N.; Sparks, S. E.; Casida, J. E. Phosphoacetylcholinesterase: toxicity of phosphorus oxychloride to mammals and insects that can be attributed to selective phosphorylation of acetylcholinesterase by phosphorodichloridic acid. *Chem Res Toxicol*, 2000, **13**(7), 652–657.

59. Sese, B. T.; Grant, A.; Reid, B. J. Toxicity of polycyclic aromatic hydrocarbons to the nematode *Caenorhabditis elegans*. *J Toxicol Environ Health, Part A*, 2009, **72**(19), 1168–1180.

60. Cavalieri, E.; Rogan, E.; Cremonesi, P.; Higginbotham, S.; et al. Tumorigenicity of 6-halogenated derivatives of benzo[a]pyrene in mouse skin and rat mammary gland. *J Cancer Res Clin Oncol*, 1988, **114**(1), 10–15.

61. Grandjean, P.; Andersen, E. W.; Budtz-Jørgensen, E.; Nielsen, F.; et al. Serum vaccine antibody concentrations in children exposed to perfluorinated compounds. *JAMA*, 2012, **307**(4), 391–397.

62. Eriksen, K. T.; Raaschou-Nielsen, O.; Sørensen, M.; Roursgaard, M.; et al. Genotoxic potential of the perfluorinated chemicals PFOA, PFOS, PFBS, PFNA and PFHxA in human HepG2 cells. *Mutat Res—Genet Toxicol Environ Mutagen*, 2010, **700**(1), 39–43.

63. 3M Company. 3M™ fluorosurfactant FC-4432: 98-0212-3045-7. https://multimedia.3m.com/mws/mediawebserver?mwsId=SSSSSuUn_zu8l00x482xnxmUov70k17zHvu9lxtD7SSSSSS (accessed January 15).

64. O'Brien, P. J. Molecular mechanisms of quinone cytotoxicity. *Chem Biol Interact*, 1991, **80**(1), 1–41.

65. Wang, X.; Thomas, B.; Sachdeva, R.; Arterburn, L.; et al. Mechanism of arylating quinone toxicity involving Michael adduct formation and induction of endoplasmic reticulum stress. *P Natl Acad Sci USA*, 2006, **103**(10), 3604–3609.

66. Osimitz, T. G.; Nelson, J. L. Understanding mechanisms of metabolic transformations as a tool for designing safer chemicals. In: Anastas, P. T.; Boethling, R.; Voutchkova, A. (Eds.), *Handbook of Green Chemistry*, Wiley-VCH: Weinheim, Germany, 2012, pp. 47–76.

# Chapter 8

# Case Studies

In this chapter, we will describe several examples where traditional technologies and molecules, often highly hazardous to human health and the environment, were replaced by safer, benign, and environmentally friendly alternatives. These range from biodegradable flame retardants, chelants, and surfactants; to better drugs, safer insecticides, and antifoulants, among others. For each of the following examples, we ask four questions that are answered in a few key paragraphs:

(1) What came before?—describing old technology/process/molecule
(2) What is this new thing?—describing safer alternative
(3) Why is it superior?—describing safety, performance, cost improvements/benefits
(4) How, from the molecular design perspective, was this accomplished?

## Example I: Safer Flame Retardants

### What Came Before?

Due to its hazardous potential, the use of decaBDE and other PBDEs was restricted by EuropeanRoHS Directive[1] and additionally, the

*First Do No Harm: A Chemist's Guide to Molecular Design for Reduced Hazard*
Predrag V. Petrovic and Paul T. Anastas

ISBN 978-981-4968-59-1 (Hardcover), 978-1-003-35964-7 (eBook)
www.jennystanford.com

U.S. EPA has pressured manufacturers to stop production. Thus, the use of these flame retardants is being phased out globally. Commercially available alternative halogen-free flame retardants that are currently used are based on phosphorus compounds. For example, *N*-methylol phosphonopropionamide derivatives and hydroxymethylphosphonium salts are used in coatings of the cotton fabrics in commercial flame-resistant fabrics Pyrovatex® and Proban®.[2] However, there are still drawbacks with using these compounds, as dangerous formaldehyde is released during their application and long-term use.[3] Also, the EPA's Committee on the Design and Evaluation of Safer Chemical substitutions described in their report[4] that several assessed alternatives to the decaBDE shown on Fig. 8.1, each had their own disadvantages if they are to be widely used. They either have not met U.S. flammability requirements with non-halogenated retardants, did not meet ecolabel criteria, or the material cost would have been very high.

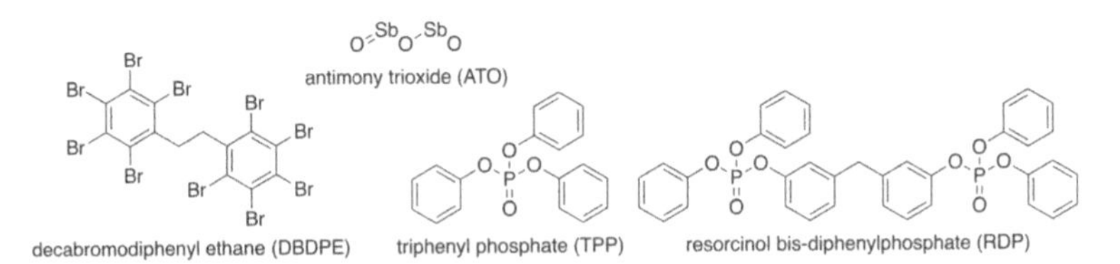


**Figure 8.1** Commonly used flame retardants.

## What Is This New Thing?

As an alternative to traditional flame retardants, in 1993 PYROCOOL Technologies, Inc. developed PYROCOOL F.E.F. (Fire Extinguishing Foam)—a cooling and fire extinguishing agent.[5] Their formulation contains highly biodegradable surfactants which were designed to be used in small quantities. To avoid long-term problems for health and the environment often connected with traditional flame retardants based on the poly/perfluorinated hydrocarbons, the company decided to exclude fluorosurfactants and glycol ethers. As a result, PYROCOOL F.E.F formulation contains biodegradable anionic, nonionic, and amphoteric surfactants at a very low mixing ratio with water (0.4 %). Initial performance testing demonstrated high efficacy against a broad range of combustibles, and today this product is, among other similar products, widely used at numerous airports across the world.[6]

While the use of bulk flame retardants is mostly very effective, interest in the use of surface treatment of materials in order to limit the flame-retardant reactivity at the exterior of the material has increased in recent years. This way, the amount of the additive is reduced, while the desirable bulk properties of the treated material are retained. To achieve this, flame-retardant nanocoatings are being deposited using layer-by-layer (LbL) technology to fabric, foam and other polymer substrates, or directly impregnated into the textile materials. With the advancement of 3D printing and the fact that it is becoming widely available for everyday use, the flammability of the filaments used in printing and the potential for serious hazards have also become apparent. To combat this, polylactic acid (PLA) is being mixed with other flame-retarding molecules to achieve self-extinguishing properties.[7] Grunlan's group developed flame retardants in the form of nanocoatings and an environmentally relatively benign process for their LbL deposition on textile and foam materials.[8] Additionally, they have developed aqueous polyelectrolyte complexes (PECs) consisting of polyvinylamine and poly(sodium phosphate) mixed with PLA for self-extinguishing filaments used in the 3D printing and the production of self-extinguishing textiles (Fig. 8.2).[7b,9]

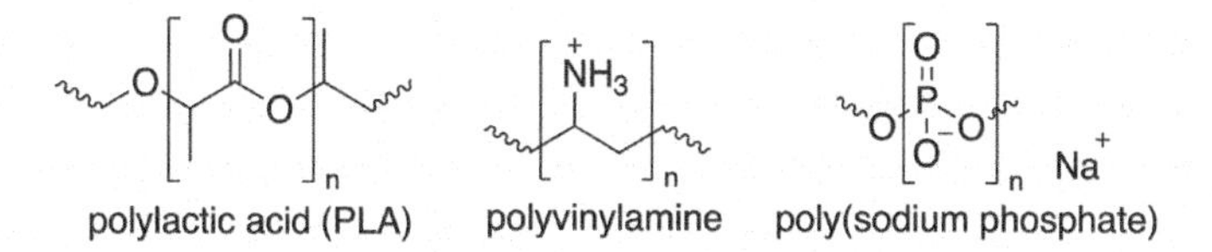


**Figure 8.2** Components of Grunlan's self-extinguishing filaments used in the 3D printing.[7b]

Another greener approach relies on the use of biomacromolecules and bio-sourced products, especially for protecting natural and synthetic textiles from flame.[10] These molecules, i.e., whey protein, caseins, hydrophobins, chitin, cyclodextrin, DNA, phytic acid, and various bio-sourced product such as tannin and lignin, can be applied by impregnation or layer-by-layer technique. Fire retardancy of these molecules is correlated with their structure and composition, the type of treated substrate, and achieved final dry add-on.

## Why Is It Superior?

PYROCOOL F.E.F. technology demonstrates dramatic enhancement of water's fire extinguishment effectiveness, while at the same time removing the environmental impact of traditional fire extinguishing foams. Because only 0.4 % of biodegradable surfactant components are mixed with water, numerous advantages are provided compared to traditional polyfluorinated flame retardants. These range from reduced acquisition costs, where smaller amounts of the product can be used to achieve the same level of effectiveness; storage requirements, container disposal efforts, and shipping costs are reduced as well; and there are no after-use costs for environmental clean-up due to the biodegradability of the product.

Grunlan et al. developed successful flame-retardant (FR) coating systems for polyester-cotton blend,[8b] a 3 mm thick polystyrene plate,[8a] and polyurethane foam.[8c] One of the advantages of using LbL FR coatings on all thermoplastic-based materials is the prevention of melt drippings. Nanoparticle-based and clay coatings protect treated materials from degradation by acting as thermally insulating barriers. Water-based PECs are applied to cotton fabric in a two-step process, making it self-extinguishing.[9a] A polyphosphate in this coating promotes char formation, which creates a thermal shield that protects the treated fabric from fire. For nylon-cotton blend fabric, additional steps include a coating of water-based halogen-free melamine which generates melamine polyphosphate in situ.[9b] The PEC coating additionally demonstrated good durability during tests in home-laundering cycles even in boiling water with preserved FR activity.[9c] This proves several advantages over traditional FR coatings; firstly, the ability to prevent ignition of different fabric types in very few processing steps; secondly, the use of environmentally benign components; and finally, relatively small increases in weight. Grunlan's group successfully incorporated PECs at a 25 wt% loading into a common PLA 3D-printing filament.[7b] This was made possible due to the plasticizing effect of water on the complex, which enables good mixing with PLA, resulting in extrusion of high-quality filament. The physical and thermal properties of this composite filament do not differ much from commercial PLA filament. Compared to the commercial PLA, which normally degrades, PLA–PEC filament demonstrated good thermal stability and demonstrated self-extinguishing behavior while exposed to open flame.

Compared to traditional flame retardants, biomacromolecules and bio-sourced extracts have a much lower environmental impact without volatile organic carbon (VOC) species release. This is additionally supported by the good protection of textile materials from heat influx and flame when biomacromolecules are applied on the surface. They act by limiting the transfer of heat, fuel, and oxygen between the material and the flame with flame retardancy on par with some phosphorus-based flame retardants. Some of the bio-soured molecules (such as cyclodextrins) are water-soluble, not toxic, and generally not expensive.[11] Another advantage comes from the possibility to acquire these molecules from agricultural/food industry sources. At this time there are still some caveats to using these molecules on a larger scale since their application was demonstrated only at the lab-scale level. One of the problems is the poor durability of the biomacromolecular coating when the textile materials are exposed to washing with water at any high temperature cycles. Another issue is that upon application of these biomacromolecules, the material becomes stiffer which limits its application in fabrics for clothing affecting consumer comfort level, and additionally some biomacromolecules (e.g., DNA) are still quite expensive to produce in large quantities. However, there is a lot of research work being done to overcome these limitations.

## How, From the Molecular Design Perspective, Was This Accomplished?

As we discussed in previous chapters, it became evident that halogenated flame retardants (such as decaBDE) posed a huge human health hazard, and thus their manufacture and use was discouraged and even banned in some countries. All of the new approaches we described above relate closely to the design suggestion we presented in Section 7.3 of this book.

PYROCOOL Technologies' approach of using biodegradable surfactants instead of traditional polyfluorinated hydrocarbons is also directly related to the design rules described in Section 7.10. Biodegradation tests conducted by independent laboratories demonstrated rapid biodegradation of fire extinguishing foam components in PYROCOOL F.E.F. and concluded that over 95% of the foam would be biodegraded within 20 days of atmospheric exposure.[5] This shows that the selection of nonionic, anionic, and amphoteric components in the formulation enabled higher bioavailability to the bacterial communities. At the same time, aquatic toxicity studies revealed that the $LC_{50}$ levels after 48 hours are very large if the 0.4% mixture with water was used (12500 mg/L), while 86% of potentially toxic components would biodegrade within 48 hours.[5]

The use of natural and environmentally benign kaolin (clay) and other nanoparticle-based coatings for FR treatment of materials, removed the inherent hazard of traditional organohalogen flame retardants.[8b,c,e] This is in good agreement with our design suggestions to limit the use of molecules with more than three halogen atoms, and increase the potential for biodegradation, while at the same time limiting the toxic mechanism of action. Additionally, the use of the PEC complex which is water-soluble and consists of environmentally more benign chemicals compared to organohalogens demonstrates thoughtful chemical design. The combination of eco-friendly PLA with the PEC complex for the filaments used in 3D printing shows the thoughtful design for the novel application of environmentally benign flame retardant.

The use of biomacromolecules as flame retardants directly eliminates the toxic mechanism of action and persistence associated with organohalides. Biomacromolecules offer several advantages over traditional flame retardants. Design suggestions stated in Section 7.10 for reduced persistence state that larger molecules are generally less biodegradable, however, even though the biomacromolecules are very large, they possess numerous structural features and functional groups that enable their biodegradation. Additionally, biodegrading organisms evolved to decompose these molecules easily. The large $M_w$ and size of the biomacromolecules also disfavor the typical toxic mechanism of action that much smaller organohalides usually undergo.

## Example II: Safer Type 2 Diabetes Drugs

### What Came Before?

Some of the most successful diabetes mellitus type 2 drugs based on thioazolidinedione functional group (see Fig. 8.3) have been removed from the market due to the severe side effects that caused harm to the patients after being prescribed for longer periods of time.[12] Troglitazone (Rezulin) was discovered to cause hepatotoxicity and pioglitazone (Actos) caused an increased risk of bladder cancer, while rosiglitazone (Avandia) shown increased toxicity to the cardiovascular system.[4] These drugs increase the cell dependence on carbohydrates (particularly glucose) oxidation for cellular energy generation by increasing deposition of fatty acids in the adipocytes, which also decreases cell resistance towards insulin. This is achieved through the activation of specific nuclear receptors—peroxisome proliferator-activated receptors (PPARs) whose natural ligands are free fatty acids. In silico studies on hepatoxicity of troglitazone revealed that it is related to the apoptotic effect due to the presence of the chromane ring which is missing in pioglitazone and rosiglitazone. Both pioglitazone and rosiglitazone have similar physicochemical characteristics (melting point, boiling point, vapor pressure) to troglitazone; however, the computational assessment show they should be much more water-soluble. This can potentially explain why these two drugs have exhibited hepatoxicity, but not their other toxic behavior.[4]

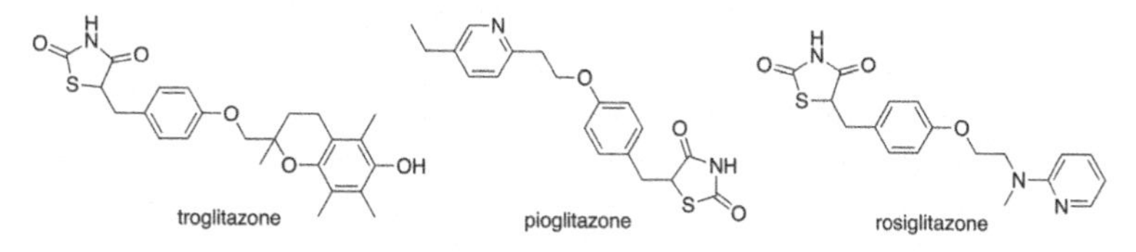


**Figure 8.3** Structures of most successful type 2 diabetes drugs.

## What Is This New Thing?

As the research for alternative antidiabetic drugs was developing, it was discovered that patients with diabetes show hyperglycemia due to an increased glucose renal reabsorption capacity. This was caused by overexpression of sodium–glucose cotransporter type 2 (SGLT-2) proteins in the kidneys that are responsible for glucose reabsorption from the glomerular filtrate to the plasma.[13] This provided an opportunity for developing a drug that would specifically target this protein and inhibit its function—this resulted in the approval of the first antidiabetic SGLT-2 inhibitor, dapagliflozin (Forxiga® or

Farxiga®, Bristol-Myers Squibb & AstraZeneca), in 2012 in Europe;[14] the U.S. Food and Drug Administration, approved canagliflozin (Invokana®, Janssen Pharmaceutical) in 2013 (Fig. 8.4).

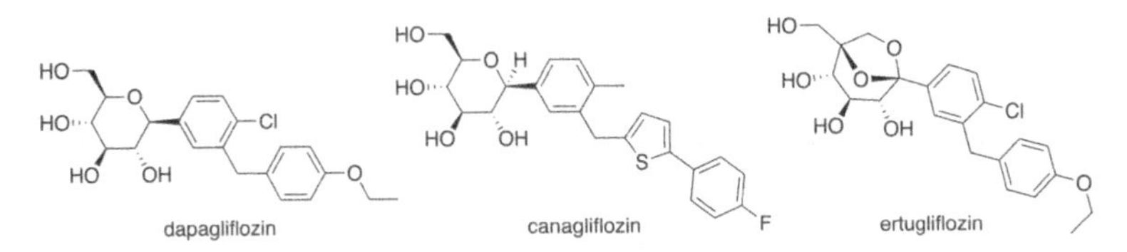


**Figure 8.4** Sodium–glucose cotransporter type-2 (SGLT-2) inhibitors.

## Why Is It Superior?

Toxicity studies for glitazones show that troglitazone exhibits low acute oral mammalian toxicity ($LD_{50}$ > 2000 mg/kg), rosiglitazone high toxicity ($LD_{50}$ = 300 mg/kg), and pioglitazone very high toxicity ($LD_{50}$ = 181 mg/kg).[4] All three of these drugs have also demonstrated increased carcinogenicity in mice and have been categorized as moderate carcinogens.[4]

The new class of type 2 diabetes drugs represents an alternative approach for controlling glucose blood levels that do not involve direct insulin intervention. Selective inhibition of the SGLT-2 protein brings additional benefits since it regulates blood pressure and reduces body weight, which is often an accompanying issue for patients suffering from type 2 diabetes. As such, these drugs have been approved for use along with other antidiabetic agents used across the world. Additionally, since these SGLT-2 inhibitors have been developed using a careful molecular design by modifying specific pharmacophore subunits, there is a lot of room for improvements toward higher affinity and the development of next-generation inhibitors.

This becomes more important considering that the current generation of inhibitors exhibits adverse effects related to increased genital and urinary infections.[13,15] Another issue is the increased risk of diabetic ketoacidosis—elevated ketone bodies in the blood causing a severe clinical condition.[16] Based on the results obtained by ongoing clinical trials, new findings will enable proper assessment of the risk-benefit relation for these inhibitors, which will, in turn, enable better and safer drugs.

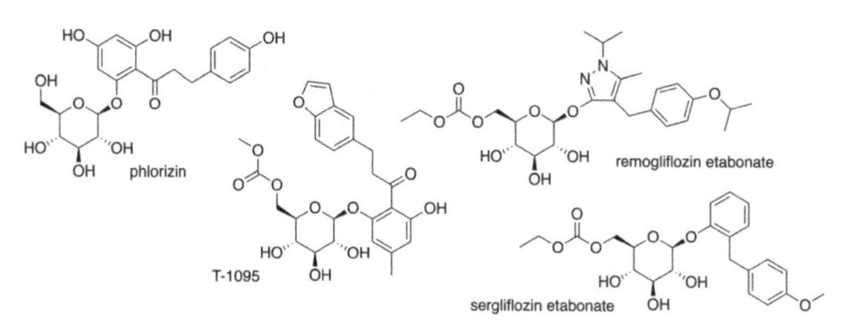


**Figure 8.5** 1st generation SGLT-2 inhibitors.

## How, From the Molecular Design Perspective, Was This Accomplished?

In the 19th century a natural O-glycosidic product—phlorizin (Fig. 8.5), was isolated from the root of an apple tree for the first time. It was discovered that this compound causes glycosuria and polyuria, i.e., mimics characteristic clinical symptoms of diabetes.[14b,17] Its function was connected with the inhibition of SGLT proteins, and the potential for diabetes treatment was investigated. However, this molecule was a nonselective inhibitor, caused unwanted adverse effects in the GI tract, and was susceptible to hydrolysis of O-glycosidic bond by β-glucosidase releasing nephrotoxic dihydrochalcone phloretin.[14b,17b,18]

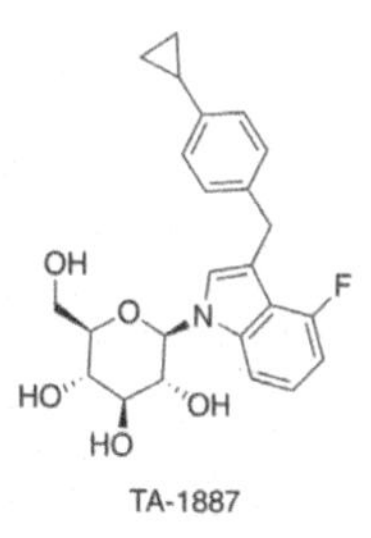


**Figure 8.6** 3rd generation SGTL-2 inhibitor.

First structural analogs were developed by Tsujihara et al. from the Japanese pharmaceutical company Tanabe Seiyaku with the goal of reducing the toxicity of the aforementioned metabolite, while at the same time preserving the SGTL-2 inhibition effect (Fig. 8.5).[19] They have employed a careful molecular design strategy by iteratively changing specific pharmacophores of the molecule which coincides with the general design suggestions to reduce toxicity we described in Sections 7.2 and 7.13. Other pharmaceutical companies joined the effort and developed other O-glycoside analogs of phlorizin, such as sergliflozin etabonate[20] and remogliflozin etabonate[21] by GlaxoSmithKline. These, along with T-1095 from Tanabe Seiyaku represent the first generation of SGLT-2 inhibitors. However, all of them failed to show satisfactory performance and development was stopped after clinical trials on human subjects.

Second-generation inhibitors are C-glycosidic analogs of phlorizin, designed to avoid hydrolysis of O-glycosidic analogs in the GI tract (Fig. 8.4). The first of these, dapagliflozin (Forxiga®) was developed by Bristol-Myers Squibb and AstraZeneca while canagliflozin (Invokana®) was developed by Mitsubishi Tanabe Pharma. Other C-glycosidic analogs such as ertugliflozin (Steglatro®) by Merck & Co. have been approved for type 2 diabetes treatment.[22]

Third-generation inhibitors are N-glycosidic analogs of phlorizin, where the drug development involved the introduction of an amino bridge that was later cyclized to improve stability of the drug candidate T-1887 (Fig. 8.6). However, these molecules have not yet entered clinical trials.

## Example III: Safer Immunomodulatory Agents

### What Came Before?

Thalidomide (Fig. 8.8) is a sedative drug, first synthesized in 1954 in Western Germany by Chemie Grünentha. It was soon discovered that thalidomide could be used as a seemingly nontoxic replacement for barbiturates, and could help relieve nausea during pregnancy. However, after some time it became evident that this drug causes severe irreversible damage to the fetus, and as a result, thousands of children were born with malformed or missing limbs. This teratogenic effect is a consequence of the thalidomide-promoted degradation of proteins responsible for turning off the SALL4 gene that influences the development of the fetus. It was noticed that, since this molecule is a racemic mixture that cannot be completely separated, there is a difference in activity between R- and S-enantiomers. While S- isomer has been identified as the one most probably causing mutagenic effects, the R- isomer was believed to have some beneficial therapeutic effects; however, this was later disproved.[23] In recent decades, thalidomide has been prescribed as an immunomodulating therapeutic in treatment for multiple myeloma, a cancer of plasma cells. However, in vitro studies show that the two enantiomer forms can racemize which causes serious side effects including fatigue, painful nerve damage (neuropathy), and the formation of serious blood clots.

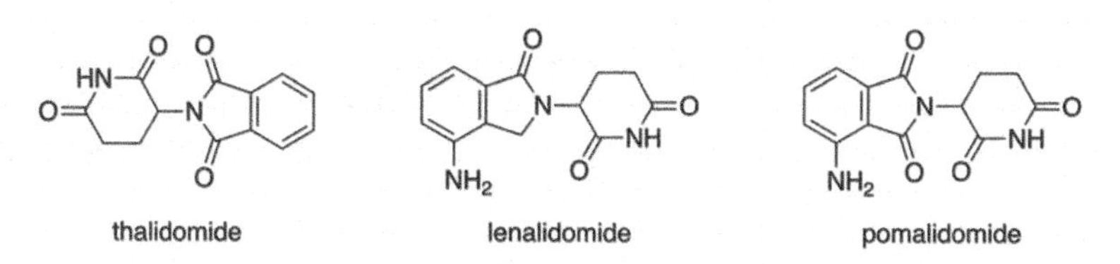


**Figure 8.7** Immunomodulatory agents used for the treatment of multiple myeloma.

## What Is This New Thing?

Alternative immunomodulatory agents lenalidomide and pomalidomide are structurally similar to thalidomide and often used as replacement therapeutics (Fig. 8.8).

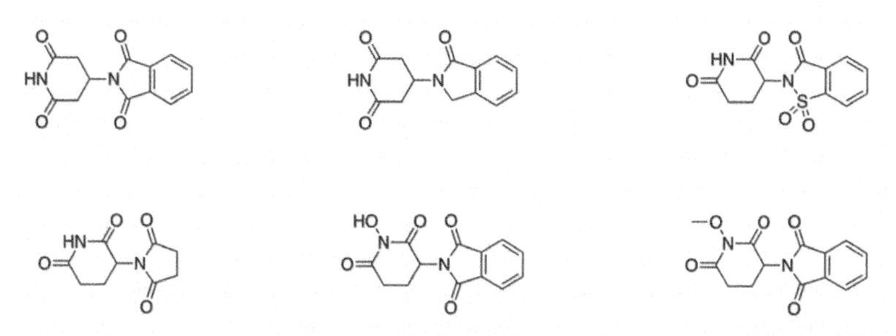

**Figure 8.8** Structural modifications to thalidomide that removed teratogenic effect.

Lenalidomide (Revlimid®)[24] alters cytokine production, regulates T cell co-stimulation, and augments the cytotoxicity of natural killer (NK) cells which causes the modulation of different immune system components. This drug exhibits direct cytotoxic effects along with the immunomodulatory behavior that is milder and more acceptable than the effects caused by thalidomide.[25]

Lenalidomide has been approved for treating multiple myeloma, myelodysplastic syndromes, and several lymphomas (follicular, mantle cell lymphoma, marginal zone lymphoma).[25]

Pomalidomide (Pomalyst®) is also used to treat multiple myeloma mostly for patients that have already received treatment with other immunomodulators. Typical side effects include low red blood cell counts (anemia) and low white blood cell counts. The risk of nerve damage is not as severe as it is with the other immunomodulating drugs, but the administration of this drug is also linked to an increased risk of blood clots. Pomalidomide is generally used in combination with other myeloma drugs such as monoclonal antibodies, alkylating agents, and proteasome inhibitors.[26]

## Why Is It Superior?

While thalidomide use as an oral sedative was abandoned due to the severe adverse effects it exhibited, its good anti-cancer properties opened a new route for the treatment of multiple myeloma. However, the limitation of the applicability during pregnancy and multiple side effects including severe neuropathy prompted the search for alternatives. Both lenalidomide and pomalidomide offer slightly better toxicological profiles with almost minimal structural modifications.[25,27] Notably, the quality of life and longevity of the patients diagnosed with multiple myeloma was increased by the introduction of these immunomodulators.[27a] That said, these new drugs still bear their share of unwanted side effects, and their application is still limited to patients that do not have other alternatives.

## How, From the Molecular Design Perspective, Was This Accomplished?

A recent study on the teratogenicity of thalidomide and structural analogues[28] revealed that both optical isomers of thalidomide act as teratogens, and due to rapid racemization and interconversion through keto-enol tautomerism, in a biological environment they form heterochiral dimer assemblies through hydrogen bonding. Both the glutarimide and phthalimide rings contribute to teratogenicity, and major structural additions to the rings interfere with the toxic activity. For example, if the linkage between the rings is changed from $\alpha$- to $\beta$-, the asymmetry and chiral nature is removed. This removes embryotoxicity and the replacement of phthalimide ring with other ring structures (e.g., succinimide, benzothiazole) or modifications on the glutarimide ring (such as the addition of a hydroxy or methoxy group on nitrogen) inactivate the teratogenic effect as well. These structural changes are compliant with the design suggestions we described in Sections 7.12 and 7.16 where problematic pharmacophores, in this case, both rings, are modified to avoid bioactivation and the toxic mechanism of action. This was in part demonstrated with thalidomide successors. While both lenalidomide and pomalidomide still exhibit various adverse effects, the structural modifications they introduced on the phthalimide ring (see Fig. 8.8) changed the behavior of the molecule, causing less severe effects. This shows the importance of a proper understanding of the potential mechanism of action for any molecule that is designed to be toxic to some extent. Targeted toxicity is only possible if careful design choices are considered, which became much easier with the advancement of in silico techniques and methodologies for testing the toxic potential of novel chemicals.

## Example IV: Safer Wood Preservatives

### What Came Before?

As a ubiquitous biological material, wood products are used in a wide variety of residential and outdoor building applications. Over time, timber in different uses may absorb water, photodegrade, and biodegrade. To prevent this, wood preservatives have been used for around 180 years to extend the lifetime of many wood products.[29] Most preservatives are biocides intended to inhibit biodegradation by insects, fungi, moss, microbes, and certain marine invertebrates.[30] Early preservatives included creosote (used to treat railroad ties) and pentachlorophenol (used for utility pole treatment),[31] but by the late 2000s, more than 95% of pressure-treated wood was preserved with chromated copper arsenate (CCA) (see Fig. 8.9).[32] One hundred and fifty million pounds of CCA wood preservatives were used in 2001.[32] CCA is an effective preservative in both terrestrial and marine environments, with chromium acting as a fixing agent, copper as a fungicide, and arsenic as an insecticide.[33] Each of these substances is toxic or ecotoxic. While chromium is reduced to Cr (III) in the fixation process,[34] Cr (VI) is a carcinogen and mutagen. Arsenic may be a carcinogen, mutagen, and teratogen. Copper, as a free ion, is toxic in the environment, particularly to aquatic organisms. The risk of children being exposed to arsenic through leaching, among other concerns due to the production, handling, and disposal of CCA-treated wood, generated demand for new preservative solutions.[32]

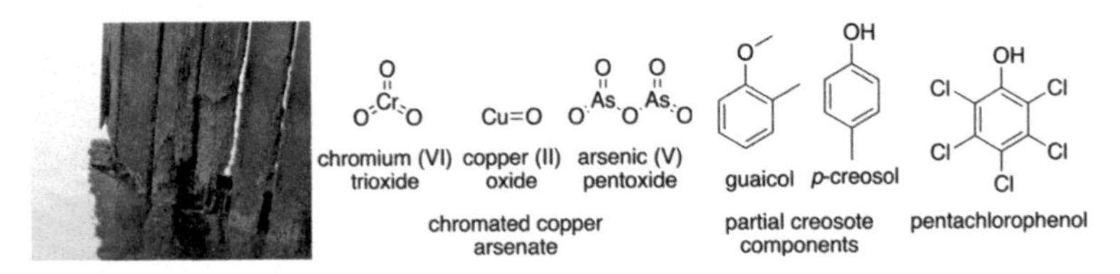


**Figure 8.9** Rotting wood[35] and earlywood preservatives.

## What Is This New Thing?

In 1996, Chemical Specialties, Inc. (now Viance) commercialized an alkaline copper quaternary (ACQ) wood preservative (Fig. 8.10).[32] A bivalent copper complex and quaternary ammonium species (quat) dissolved in ethanolamine or ammonia form the basis of the ACQ technology.[32] In this system, copper continues to function as a fungicide solubilized by a readily available amine. The quat acts as an additional fungicide and insecticide, as well as an inhibitor of wood molds and stain-causing organisms.[36] Quats are leach-resistant due to an ion-exchange process that replaces the formulating anion with acidic functional groups in the wood.[36] Over 100 million board feet of preserved wood were created using ACQ treatment by 2001, just five years after the products' release.[32] In addition to ACQ systems, recent innovations are focusing on using antioxidants and chelators to disrupt biodegradation processes in a targeted manner and chemically modifying wood to make it harder to degrade.

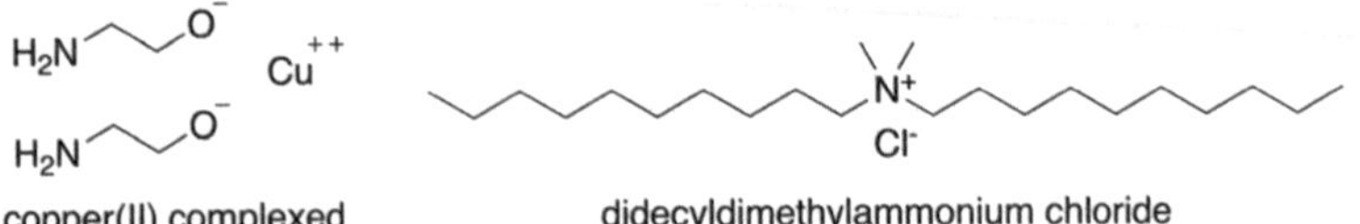


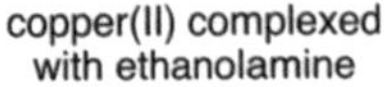


**Figure 8.10** Basic ACQ preservative.

## Why Is It Superior?

The replacement of CCA wood preservatives with ACQ-based formulations resulted in a wood protectant that is superior across multiple endpoints and lifecycle stages. First, ACQ technology removes the risk of cancer and reproductive harm from all lifecycle phases of treated wood products. In 2002 the EPA predicted that Chemical Specialties' ACQ Preserve® would eliminate the use of 64 million pounds of chromium VI and drastically reduce the quantity of arsenic (44 million pounds a year around 2001) used for wood preservation.[32] Removing these metals from the product eliminates health risks during product manufacture (formulation and application), use (skin contact and leaching), and disposal (controlled or uncontrolled burning). Replacing these metals, quats have been used as household and medical disinfectants for over 100 years,[37] particularly because of their low toxicity to mammals.[29]

Secondly, the use of ACQ technology in the United States removes the regulatory burdens otherwise required with CCA treatments. Because of the reduced toxicity, facilities using ACQ are no longer bound by the specific obligations of the Resource Conservation and Recovery Act (RCRA) in the United States, which imposes strict cradle to grave tracking and management.[32] The subsequent disposal of ACQ-treated wood is not covered by RCRA because it is expected to be lower risk.[32] These reduced regulatory burdens save producers both time and money.

While ACQ wood preservatives are favorable from mammalian toxicity and regulatory burden standpoint, it is important to note that this technology is not yet optimized. Copper is a potent aquatic toxin, and the presence of both copper and the amines result in increased corrosion to metal used in building structures.[38] The need for an amine to complex copper also results in a greater cost compared to CCA and the disposal of copper-rich wood is still a concern.[38]

In response to these and other problems with ACQ wood preservation, a number of other products emerged in the past decades. Instead of deterring or harming organisms that degrade wood, researchers built on the knowledge that some wood degradation mechanisms rely on free radicals and metals and therefore attempt to disrupt the metabolic pathways of degradation.[31,39] By incorporating antioxidants and chelators into existing biocide treatments to preserve wood, researchers have demonstrated an alternate path to wood treatment (Fig. 8.11).[31]

Alternatively, Eastman, Accsys, and other companies are working to make wood more resistant to degradation via chemical modification of the wood itself.[40] Acetylation, which is accomplished by reacting the wood with acetic anhydride (acetic acid is a byproduct), esterifies the hydroxyl groups in wood, increasing strength and reducing degradation (Fig. 8.11).[41] Instead of impregnating the wood with a preservative, this approach is thought to work either through structural changes preventing biodegradation or through the extensive removal of water in the wood needed for biodegradation.[40] Acetylation and the use of antioxidants and chelators may offer more resilient wood using chemicals of reduced hazard.

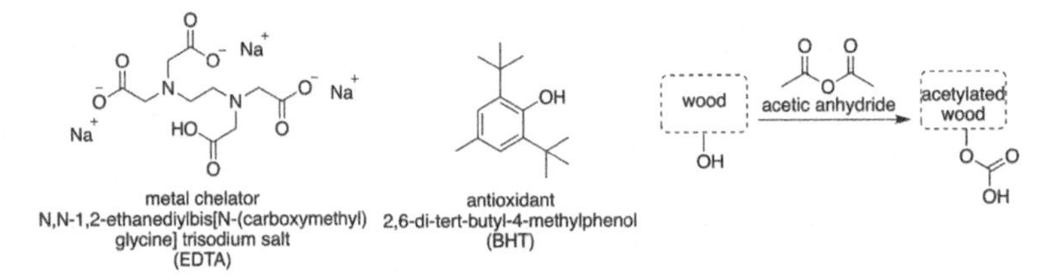


**Figure 8.11** Chelator and antioxidant used in alternative wood preservative[31] and general acetylation process.[40]

## How, From the Molecular Design Perspective, Was This Accomplished?

Applying the design rules retroactively, the first consideration for an improved wood preservative would be mammalian absorption. Didecyl dimethyl ammonium chloride (DDAC) and alkyl ($C_{12}$, $C_{14}$, $C_{16}$) dimethyl benzyl ammonium chloride (ADBAC), two quats used in ACQ formulations, are poorly absorbed via the oral and dermal route (less than 10%) in rats.[42] Their potential to bioaccumulate in humans is also assumed to be low, as evidenced by ~90% excretion in rats.[42] As noted in Chapter 7, this may be explained by DDAC and ADBAC's low log P values, 2.6 at 20 °C and 3.91 (estimated),[43] relatively high solubilities, 650 g/L at 20 °C and >1,000 mg/L,[43] and low molecular weight. By choosing a compound with a low potential for absorption and a high potential for excretion, product designers may have minimized overall bioavailability. In addition, while these compounds are moderately toxic via acute oral, dermal, and inhalation exposure, DDAC and ADBAC are not known to be developmental or reproductive toxicants and are not carcinogenic or mutagenic.[42] Selecting molecules with low bioavailability and toxicity minimized the human toxicological risks through the mitigation of chemical hazards.

As noted above, ACQ technology is not nontoxic—ecotoxicity remains an issue. The use of chelators and antioxidants to disrupt the biochemical processes may present a potential low-toxicity solution. From a design perspective, this approach does not require a mechanism of acute toxicity. This technology may be of low toxicity (EDTA is widely used in consumer products and BHT is a commonly used food preservative) by focusing on metabolic disruption instead. Similarly, wood acetylation provides another approach to safer design by removing the use of chemical impregnation altogether. Each of these approaches to wood preservation is of note because they remove the need for a toxicant.

## Example V: Safer Antifoulants

### What Came Before?

Fouling—the unintended growth of marine biota on submerged ship surfaces—is a costly problem for the shipping industry worldwide.[44] Within minutes of submersion, a film of organic compounds forms on surfaces allowing bacterial and diatom colonization.[45] In a matter of weeks, these organisms give way to protozoa, algae, mollusks, crustaceans, and others.[45] Costing an estimated 3 billion USD in 1995, these organisms increase the drag on a ship's hull, requiring additional fuel to compensate for inefficient movement.[44] Organotin antifoulants, particularly tributyl tin oxide, were widely used since the 1970s as biocides in marine paint formulations (Fig. 8.12).[44,46] While optimized organotin formulations provided up to five years of antifouling activity,[45] these substances were acutely toxic, caused malformations in shellfish, bioaccumulated, decreased the reproductive rates for certain species,[44] and caused the expression of male reproductive organs in female shellfish at concentrations as low as one nanogram per liter.[46] The use of tributyltin paints in quantities of around 136,000 kilograms per year in the late 1980s[45] coincided with regulatory restrictions like the Organotin Antifoulant Paint Control Act of 1988 and an EPA-mandated search for new antifoulants.[44]

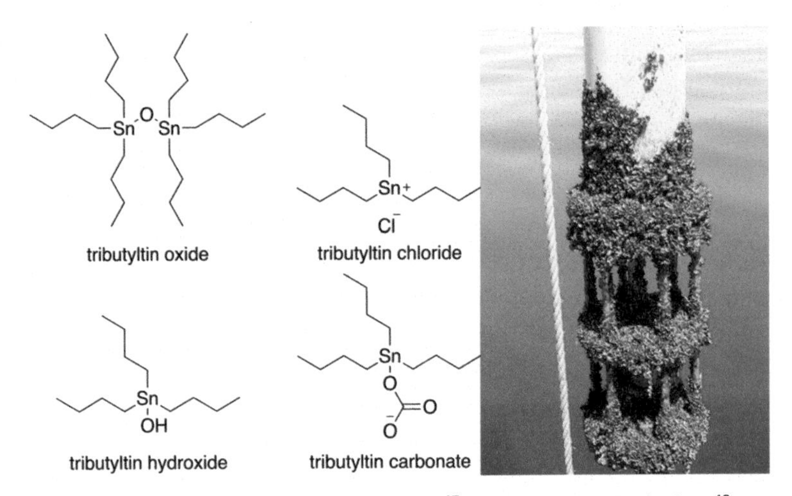


**Figure 8.12** Tributyltin species in seawater[47] and freshwater biofouling.[48]

## What Is This New Thing?

Obtaining the first new antifoulant registration from the EPA in more than a decade, Rohm and Hass Company's Sea-Nine™, emerged as an environmentally preferable alternative to organotin antifoulants.[44] Sea-Nine™, or 4,5-dichloro-2-n-octyl-4-isothiazolin-3-one (DCOIT), was one of over 140 members of the isothiazol-3-one class screened during development (Fig. 8.13).[49] The isothiazol-3-one class is known for antifungal and antibacterial properties, with related products serving as common preservatives.[50] These chemicals are known to diffuse across cell walls and membranes of fungi and bacteria whereupon the electrophilic sulfur in the S-N bond reacts with biological thiols like cysteine and glutathione.[50] As previously described, binding to macromolecules inhibits critical cell functions and causes cell death. DCOIT is biologically active against diatoms, algae, bacteria, and barnacles.[51] Sea-Nine™ is used on a global scale, and the development of this safer antifoulant garnered the 1996 Presidential Green Chemistry Challenge.[44]

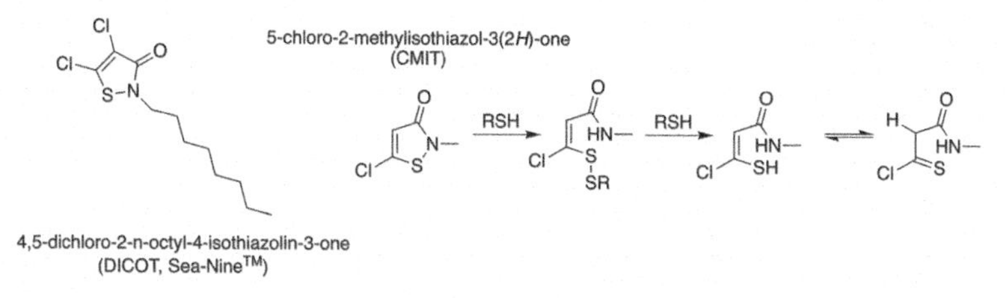

**Figure 8.13** DICOT and proposed CMIT reaction with endogenous thiols.[52]

## Why Is It Superior?

Sea-Nine™ was designated as a preferred antifoulant because of its rapid bio- and photodegradation, low potential for bioaccumulation, and minimal chronic toxicity. These improvements make DICOT superior in the use and end-of-use phase, though it remains acutely toxic to humans and therefore still a manufacturing risk.[53] Willingham and Jacobson, the chemists behind this antifoulant, conducted a battery of tests to compare DICOT to tributyltin (Fig. 8.14).[51] Initial tests indicated that DICOT's half-life was less than 24 hours in natural seawater, and less than 1 hour in a seawater and sediment microcosm, while tributyltin's respective half-lives were between several weeks to months and six to nine months.[51] In photolysis studies over a range of conditions, DICOT's half-life was between 6.8 and 18.0 days in sunlight and between 14.4 and 79.7 days in the dark.[54] At 25 °C, DICOT's hydrolytic half-life was between 25 and 28 days.[54] A bioaccumulation study in bluegill fish found rapid degradation in fish and a bioconcentration factor of up to 1,200 in fish viscera,[55] whereas tributyltin's highest reported bioconcentration factor was 10,000.[51] Based on these metrics, it is clear that Sea-Nine™ presented a marked improvement over the existing organotin antifoulants.

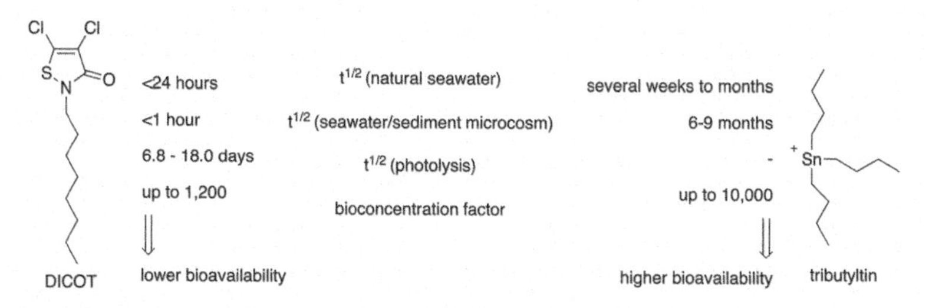


**Figure 8.14** Comparison between DICOT and Tributyltin.[51,54,55]

Despite its superiority compared to organotin species, Sea-Nine™ is not a perfect solution to fouling. While initial aerobic and anaerobic biodegradation studies in seawater indicated a biodegradation half-life of less than one hour,[51] varied conditions have produced half-lives up to 13 days.[54] The potential for the prolonged existence of DICOT in seawater may explain environmental detections of DICOT up to 3700 ng/L near areas of heavy boat traffic.[54] DICOT has been demonstrated to act as an endocrine disruptor in some species and detrimentally impact some species' reproduction.[54] Worst-case scenario calculations show that it could endanger many marine biotas.[54] Additional antifouling agents are being developed which may have reduced endocrine disruption and lower relative toxicity.[56] A number of other antifouling strategies have been proposed but have yet to be developed, such as the enzymatic cleavage of biofouling organisms' adhesive structures, the disruption of the metabolic processes of biofoulers, and chemotaxis to hold bacteria at a distance, possibly decreasing their adhesion.

### How, From the Molecular Design Perspective, Was This Accomplished?

The development of Sea-Nine™ focused on the design of a biodegradable and minimally bioavailable compound. Willingham and Jacobson approached the design of Sea-Nine™ intending to minimize adverse consequences by reducing marine exposure to the antifoulant.[49] The two justified this decision due to their belief that it would be very challenging to make a chemical effective against a broad range of organisms that would be nontoxic to anatomically similar organisms.[49] Understanding that the ADME properties of related marine organisms are likely to be similar, a selectively toxic chemical would require extensive pharmacodynamic knowledge across many species. Even then, it was not guaranteed that a sufficiently unique biological target would allow for discrimination between target and non-target organisms. Thus, DICOT developers focused on designs for rapid degradation (Section 7.10) and partitioning into the sediment to reduce bioavailability.[49]

The isothiazolin-3-one class of compounds was selected due to known activity against aquatic organisms, rapid biodegradation, and strong affinities for sediment.[49] Following tests for biological activity within the screening model, compounds were evaluated for water solubility (Fig. 8.15).[49] As indicated in the molecular design pyramid, controlling solubility influences bioavailability. In this context, decreasing solubility (which was aided by chlorination at the 4 and 5 positions)[49] was intended to both reduce leaching from the paint matrix and increase partitioning into the sediment (adsorbed DICOT was assumed not to be bioavailable).[55] Thus, decreased solubility aimed to reduce the source (leaching behavior) and increase the sink (sediment binding), resulting in an overall lower concentration. While DICOT's log P (between 2.8 and 6.4)[54] is well within the range for absorption as indicated in Section 7.5, and measured bioaccumulation factors range up to 1100, analysis of fish tissue suggested that fish metabolism proceeds via ring cleavage and subsequent incorporation into endogenous protein.[55] This may be in part explained by the suggestions for detoxification in Section 7.13 because the labile 1,2-thiazole moiety is labile. As an overall strategy, the reliance on rapid biodegradation and solubility manipulation to minimize environmental concentrations presents an example of selectively applied design rules.

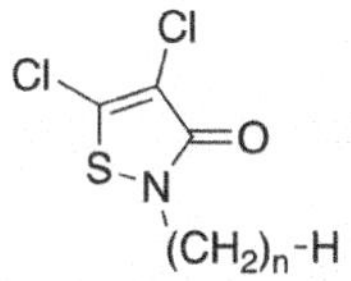

| n | bioactivity against alage | bioactivity against barnacles | estimated solubility (ppm) | |
|---|---|---|---|---|
| 4 | ✓ | X | 1000 | |
| 6 | ✓ | ✓ | 120 | |
| 8 | ✓ | ✓ | 2 | DICOT |
| 10 | ✓ | X | 1 | |

**Figure 8.15** Activity and solubility optimization for 4,5-dichloro-isothaizolin-3-one species.[49]

## Example VI: Safer Surfactants

### What Came Before?

Surfactants are organic substances where each molecule contains a strong hydrophilic group linked to a strong hydrophobic one. Because of these features, a group of such molecules will naturally aggregate on the interface between two phases—an aqueous medium and another phase that can be air, some oily liquid, or a particle. As such, they act as foaming agents, emulsifiers, or provide particle suspension.[57] Thus, they have found wide applications in numerous industries (e.g., agriculture, cosmetics, chemical, pharmacy, etc.). As a result, these industries release large quantities of a wide range of surfactants with the wastewater from their facilities (Fig. 8.16). Once released surfactants enter bodies of water where they can persist causing serious environmental problems and endangering numerous aquatic organisms.[58]

Figure 8.16 Foam on a waterbody in Freemont, California, 1972.[59]

Some of the most commonly used anionic surfactants are tetrapropylene alkylbenzene sulfonate (TPBS), linear alkylbenzene sulfonate (LAS), and nonylphenol ethoxylates (NPEs) (Fig. 8.17).[60] The toxicity of surfactants is exhibited through the absorption and penetration of cell membranes of aquatic organisms,[61] as well as a tendency to bind to various biomacromolecules such as proteins,[62] peptides, or DNA.[63] The health hazard to humans is mainly exhibited through the irritation of the top layer of the sensitive skin (see Section 2.1) causing aphthous ulcers[64] and dermatitis.[65]

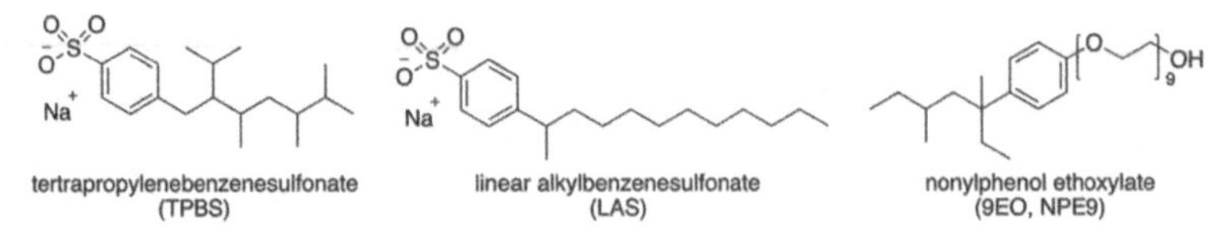

Figure 8.17 TPBS, LAS, NPE.

A surfactants' toxicity is generally correlated with its effectiveness, and how persistent they are. NPEs are nonionic surfactants that are often used in cleaning products, degreasers, emulsifiers, cosmetics, paints, etc. Due to their high performance and cost-effectiveness, they are considered "workhorse" surfactants.[66] These surfactants biodegrade slowly, and in the process, generate toxic nonylphenols.[67]

## What Is This New Thing?

TPBS surfactant was found to be not very biodegradable, causing excessive foaming in sewage treatment plants and receiving rivers with concentrations as high as 2 mg/L. This caused significant pressure from the public and government to address the issue. The cause of resistance to biodegradation was narrowed to alkyl chain branching (as we described in Section 3.8), and relatively quickly an alternative, LAS, was introduced that was produced using molecular sieves to obtain mainly linear alkanes from petroleum.[60]

As an answer to the persistence and toxicity of NPEs, companies such as Dow and BASF have been developing safer surfactants based on the use of alcohol ethoxylates (AEs). These surfactants were designed to meet the DfE Criteria for Safer Surfactants[68] and can be found in the CleanGredients database (www.cleangredients.org). Dow's ECOSURF-EH line and BASF's Inoterra® line are nonionic surfactant systems that are composed of both branched and linear ethoxylated alcohols that have similar performance to NPE surfactants.[66]

Greener surfactants have been in development as a result of increased demand for products that are safer, biocompatible, and biodegradable (Fig. 8.18). They are defined as amphiphilic molecules that are synthesized from renewable raw materials or obtained from nature.[69] For example, alkyl polyglucosides (APGs) are nonionic surfactants that are synthesized from glucose and fatty alcohols obtained from renewable raw materials.[70] They are a mixture of anomers, isomers, and homologues that are more biodegradable if their alkyl chain is longer. Due to their good cleaning abilities, wettability, emulsification, skin compatibility and foaming capability, APGs found their way to use in personal care products and cosmetics, cleaning products, food, etc.[70b] Dow Chemical Company produces TRITON™ CG-425 which is an APG with alkyl chain distribution of $C_8$–$C_{14}$.[71] Another example of surfactants that provide very low toxicity, good cleaning performance, wetting, and emulsification are rhamnolipid biosurfactants. These surfactants are naturally occurring glycolipids produced by microorganisms, mainly by *Pseudomonas aeruginosa*.[72] They are also readily biodegradable and do not produce toxic and persistent degradation products. In 2004, Jeneil Biosurfactant Company was selected as the winner of the Presidential Green Chemistry Challenge for small businesses for developing a bio-based rhamnolipid surfactant that can be commercially produced in a controlled fashion by fermentation using *Pseudomonas aeruginosa*, and isolated after sterilization and purification.[73]

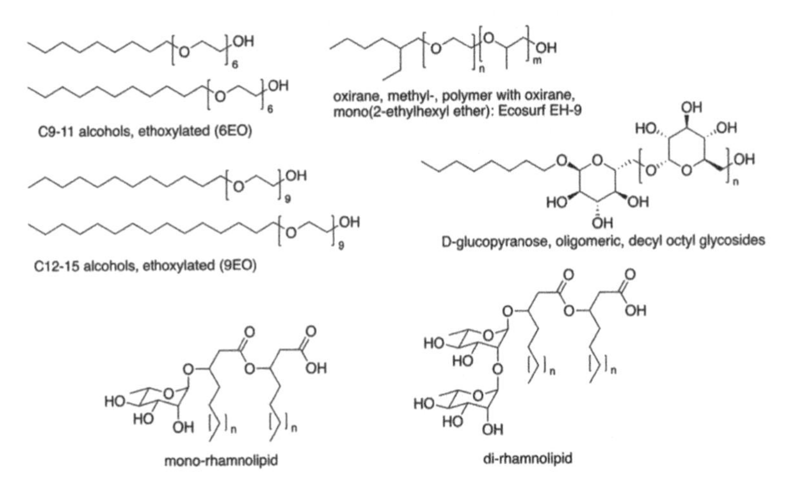


**Figure 8.18** Greener emerging surfactants.

## Why Is It Superior?

Starting from one of the first historical examples, TPBS surfactant use was abandoned even though it was very efficient and cost-effective because of the poor biodegradability and resulting toxicity in favor of LAS surfactants. Numerous studies have shown extensive biodegradation of LAS both in freshwater and seawater environment.[74] However, some portion of surfactant can be adsorbed to the solids suspended in sewage treatment facilities, and thus evade aerobic degradation which can cause unwanted environmental effects.[74]

As a valid substitution for NPEs, AEs like those produced by BASF and Dow, offer surfactants that are readily biodegradable, less sensitive to hard water, foam very little, and can clean synthetic fibers very effectively. However, some of these surfactants are still toxic, which is usually mitigated by rapid degradation in the aquatic environment.[66,75]

As more "greener" and efficient surfactants, APGs have seen increases in production in the last couple of years. They exhibit low toxicity, have good stability in a wide pH range, and show good surface tension reduction and wetting, while not causing irritation to the skin. Due to all of these properties, they are used as food and industrial emulsifiers, detergents, pharmaceutical granulating agents, and cosmetic surfactants amongst others.[70,75a] Rhamnolipid biosurfactants provide an environmentally friendly alternative to traditional surfactants. They exhibit low toxicity, emulsification, and de-emulsification activity, show anti-biofilm and anti-adhesive activity, and additionally, antibacterial activity.[72,73,76] These surfactants are produced using renewable feedstocks, requiring fewer resources, and do not require utilization of hazardous substances. They can also be used along with other synthetic surfactants, however, there are still issues with production on the industrial scale due to high production costs. Currently, they are being used in the food industry, in cosmetic products and the pharmaceutical industry, for bioremediation purposes and oil recovery, in detergents, and for agricultural purposes.[72,76]

## How, From the Molecular Design Perspective, Was This Accomplished?

The example of TPBS and LAS surfactants illustrates how alkyl chain branching can influence the biodegradability of a molecule, and that extensive branching should be avoided when designing molecules for biodegradation as we suggest in Section 7.10. Additionally, the performance of the TPBS surfactant and its use was placed before environmental considerations. In turn, this generated serious consequences and was only mitigated after the public raised concerns and chemists employed more thoughtful design strategies to create LAS.

The necessity for better-performing surfactants that will also be much safer, sparked the research efforts to find more environmentally acceptable alternatives. One of these alternatives, AEs, was a response to replace very efficient, but also persistent NPEs. AEs such as Ecosurf (Fig. 8.18), passed EPAs DfE criteria for surfactants because it exhibits low persistence but also moderate aquatic toxicity.[66] The structural features present in these ethoxylates contribute to their better biodegradability, i.e., linear instead of branched alkyl chains and ether bonds as part of ethoxylated moiety as suggested in Section 7.10; however, the presence of these features can lead to the formation of reactive degradation products that can disrupt normal cell functions and cause toxicity. As greener alternatives, bio-based and bio-sourced surfactants have gained a lot of attention. Alkyl polyglucosides offer nonionic surfactant products that are readily biodegradable due to the presence of numerous OH groups and linear alkyl chains which contribute to greater susceptibility to enzymatic hydrolysis as mentioned in Section 7.10. Rhamnolipid surfactants follow the same trend of good biodegradation potential as compared to alkyl polyglucosides. They are readily bioavailable to microorganisms possessing several structural features (see Fig. 8.18) that contribute to easier biodegradation such as the ester moiety, OH groups, and glucose unit (as suggested in Fig. 7.7).

## Example VII: Greener Chelating Agents

### What Came Before?

Chelating agents, or chelants, are organic chemicals that have two or more electron-donating atoms and can form stable complexes with metal ions. Most chelants contain some combination of amino, alkoxy, carboxyl, phosphonic acid, and phosphate functional groups that form strong coordination bonds with a metal ion (Fig. 8.19). Usually, these complexes are soluble in water and prevent metal ions from directly interacting with biological systems.[77] These chemicals found wide application in laundry detergents and for industrial cleaners, and also the cosmetic, textile, metal, paper, and photographic industries. Strong chelators can be divided into two main classes: aminopolycarboxylates and aminopolyphosphonates.[60] The most commonly used chelants, shown in Fig. 8.20, are ethylenediaminetetraacetic acid (EDTA), related diethylenetriaminepentaacetic acid (DTPA), amino tris(methylenephosphonic acid) (ATMPA), and ethylenediamine tetra(methylenephosphonic acid) (EDTMPA). Even though these substances are powerful and very effective chelants, since the

1970s they have started to raise environmental concerns due to poor biodegradability. This fact could lead to the remobilization of toxic heavy metals from sediments after their release from the wastewater treatment plants, which can cause major adverse effects to the environment.[78]

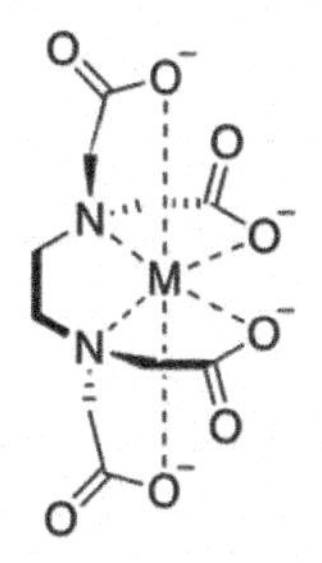

**Figure 8.19** EDTA chelation complex.

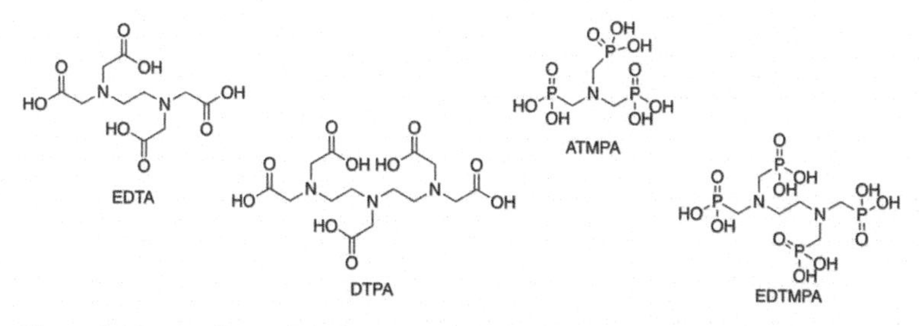


**Figure 8.20** Traditional chelating agents.

## What Is This New Thing?

Concerns about the poor biodegradability of chelants present in the market in the 1980s inspired the search for high-performing, cost-effective, and biodegradable alternatives. This resulted in the discovery of a number of lead compounds that found applications in numerous industries. Two main classes emerged above the others; polysuccinates that are based on the structure of aspartic acid, and aminocarboxylates derived from nitriloacetic acid (NTA).[77] In 2001, the Presidential Green Chemistry Challenge award was given to the Bayer Corporation,[79] which patented the first environmentally friendly manufacturing process for the commercial production of sodium iminodisuccinate (IDS) (Fig. 8.19). This molecule is readily biodegradable and nontoxic and has shown excellent chelation capabilities towards calcium, copper (II), and iron (III) ions.[79] Another greener chelant, ethylenediamine disuccinate (EDDS) is structurally similar to EDTA (Fig. 8.21). It has two chiral carbons and exists as three stereoisomers, of which only one is readily biodegradable (S,S). The other two stereoisomers biodegrade to more persistent metabolites.[80]

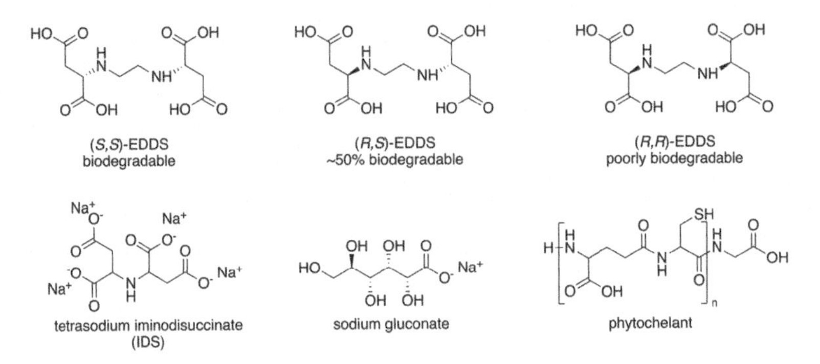


**Figure 8.21** Emerging environmentally friendly chelating agents.

One of the high-performing and environmentally friendly chelating agents that have attracted a lot of attention is sodium gluconate, a sodium salt of gluconic acid (Fig. 8.21). It functions over a wide pH range efficiently forming stable chelates with divalent and trivalent metal ions (e.g., aluminum, calcium, copper, iron).[81]

A new direction in the remediation of heavy metal pollution is employing the plants that naturally possess enzymatically synthesized peptides—phytochelatins (PC) that are known to be involved in absorbing and accumulating heavy metals. PCs consist of only three amino acids, glutamine (Glu), cysteine (Cys), and glycine (Gly) where the Glu and Cys residues are linked through a $\gamma$-carboxylamide bond.[82]

## Why Is It Superior?

Most of the widely used chelators shown in Fig. 8.20 are poorly biodegradable and very persistent in the environment.[60,78b] Additionally, some of these chelants are produced from amines, formaldehyde, and hydrogen cyanide.[79,81c] The Bayer Corporation solved this issue by developing a safer production process for IDS using maleic anhydride, ammonia, sodium hydroxide, and water as the only solvent. Ammonia formed as a side product is recycled back into the production process.[79] IDS is a readily biodegradable and environmentally benign chelator[81c,83] that found various applications as a softener and stabilizer in detergents, in photographic film processing to eliminate the precipitation of metal ions, and in agriculture to prevent, correct, and minimize crop mineral deficiencies.[79] On the other hand, only one of the three EDDS stereoisomers is readily biodegradable (S,S-stereoisomer), but it has shown superior chelating abilities in the remediation of metal pollution, in pulp and processing, and laundry detergents.[84]

Sodium gluconate is a readily biodegradable molecule that offers excellent performance as a chelating agent, corrosion inhibitor, sequestrant, humectant, and thickener which led to a wide application in the food, pharmaceutical, textile, and construction industries.[81,85] In February 2020, the U.S. EPA marked sodium gluconate as one of the 20 low-priority substances under the Toxic Substances Control Act (TSCA), which designated it as not needing risk evaluation.[81a] This was supported by a detailed report that shows that this chelant is a benign and readily biodegradable molecule.[86]

Due to increased human activities such as vehicle emissions, metal processing, toxic waste spillage, the soil around the world becomes contaminated with heavy metals; thus, the decontamination of these sites has increasingly become a necessity.[87] PC chelants offer a much safer alternative to traditional chelants such as EDTA, as they are naturally produced by plants after exposure to different metals albeit to various levels depending on the metal.[82b] Thiol (-HS) groups from cysteine residues form a chelate ring with metal ions which play a crucial role in the performance of these polypeptides. Since the gene responsible for the PC synthase enzyme activity was identified in *D. tertiolecta*, the road is open to modifying the other plants' genome so they can be utilized for remediation purposes on a larger scale.[82]

## How, From the Molecular Design Perspective, Was This Accomplished?

Even though in Section 7.7 we suggested avoiding the chelation to limit the absorption of toxic metal in the GI tract, chelates have proven to be very beneficial for numerous applications. When the Bayer Corporation developed their process for safer production of IDS,[79] they eliminated the use of toxic reactants and solvents, while at the same time producing a biodegradable and benign molecule. IDS contains a secondary N atom that contributes to easier biodegradation compared to EDTA and DTPA, which both have multiple tertiary N atoms in the structure. This is concurrent with the trend in biodegradation coefficients we presented in Fig. 3.17. Additionally, even though IDS has several stereoisomers due to the presence of chiral centers, all of them are readily biodegradable due to the presence of C-N bonds that are broken by the C-N lyase enzyme which results in the formation of fumaric and L-aspartic acids.[83b] As we mentioned in Section 7.10.1, reducing steric hindrance and size, and the conversion of tertiary amine to secondary amine, made this molecule bioavailable for enzymatic cleavage. Similarly, EDDS is a close analog to EDTA, however, it also lacks tertiary amine nitrogen atoms which contributes to better biodegradation. However, in the case of EDDS only the (S,S) isomer is readily biodegradable, which again reflects our suggestions in Section 7.10, and in 7.16 that suggest avoiding chiral centers to limit toxic mechanisms of action. Another issue that can cause a difference in the biodegradability of these chelants is the presence of metal ions.[60] As mentioned above, metal ions preferentially bind to chelants, which stabilize the molecule and make it less bioavailable to the biodegrading microorganisms.

Sodium gluconate bears a number of structural features that we suggested in Chapter 7 that contribute to its benign characteristic. It has five hydroxyl groups and a terminal carbonyl group that contributes to its chelating behavior. It is a solid substance under ambient conditions, and as a salt, it is not expected to be volatile and easily inhaled. This also contributes to the water solubility, and having a $M_w$ of 218, it has the potential to be absorbed in the lungs and GI tract. However, the estimated low $K_{ow}$ value indicates that sodium gluconate will not cross lipid membranes and bioaccumulate in fatty tissue. Experimental data indicate that this molecule is readily biodegradable both via aerobic, and ultimately anaerobic routes.[86]

## Example VIII: Greener Biochemical Pesticides

### What Came Before?

In 2009, approximately 5.6 billion pounds of pesticides were used worldwide.[88] First developed as insecticides in the 1930s and 40s,[89] organophosphates were the insecticide of choice in the United States (around 70% of the market) and many other countries until at least 2000.[90] Though the use of organophosphates in the United States declined significantly in the past two decades, these insecticides continue to be used at large volumes globally, particularly in developing nations.[90] As a chemical class, organophosphates are neurotoxins with low species selectivity.[90] Insecticides like malathion, parathion, and chlorpyrifos belong to a sub-family known as phosphorothioates, which contain a P=S bond (see Fig. 8.22). In insects as well as mammals, the P=S bond may undergo oxidative desulfuration, producing a P=O bond.[90] As shown previously in Fig. 7.35, these bioactivated molecules bond to acetylcholinesterase, which is an enzyme responsible for hydrolyzing the major neurotransmitter acetylcholine.[90] Once bound by an organophosphate, acetylcholinesterase is unable to hydrolyze acetylcholine, causing a buildup and subsequent overactivation of receptors.[90] This desirable ability to incapacitate or kill insects has resulted in widespread poisoning in non-target organisms, including humans. Organophosphates are generally of high acute toxicity via oral and dermal exposure, and the resulting "cholinergic syndrome" can cause increased sweating, tremors, nervous system effects, and even death due to respiratory failure.[90]

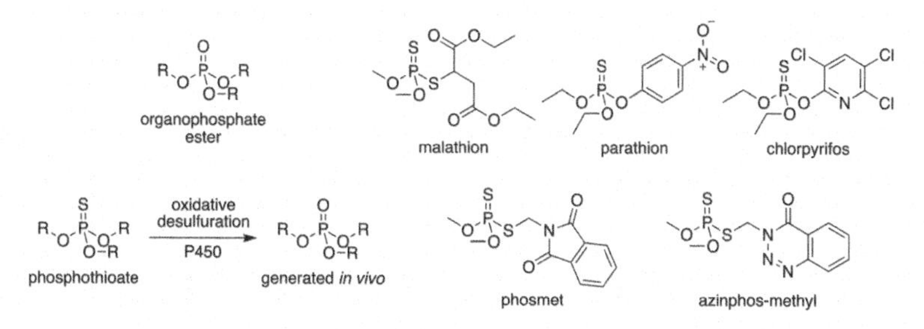


**Figure 8.22** Common organophosphates and oxidative desulfuration of phosphotioates.[90]

## What Is This New Thing?

Winner of the 1999 Presidential Green Chemistry Challenge for designing greener chemicals, Corteva™'s spinosad represents a new class of targeted insecticides (Fig. 8.23).[91] The spinosyns, of which spinosad is a member, are a class of macrocyclic lactones with a tetracyclic core and two affixed sugars.[91] Spinosyns were identified as the active ingredient produced by the *Saccharopolyspora spinosa* bacteria, which was discovered in high-volume soil screening for bioactivity.[91] Registered in more than 80 countries as of 2012,[92] spinosad, which is a mixture of spinosyns A and D, protects against caterpillars, fly pests, mosquitos,[93] and other pests when applied on trees, fruits, turf, vegetables, and other crops.[91] Spinosad works through a novel mode of action affecting neurotransmission likely via nicotinic receptors and $\gamma$-amino-butyric acid neurotransmission.[93b] Insects orally or dermally exposed to spinosad display involuntary contractions, disrupted coordination, and eventual paralysis and death (dependent on dose).[91] Because spinosad is not very effective against a number of important tree fruit pests, Corteva™ developed the related insecticide spinetoram (Fig. 8.21).[94] Using an artificial neural network to interpret structure-activity data, researchers identified more potent spinosyns J and L which comprise spinetoram.[94] Authorized in 2007,[94] spinetoram has residual activity between 4 and 6 times greater than spinosad against select pests.[93a] Honoring this achievement, spinetoram won the 2008 Presidential Green Chemistry Challenge for designing greener chemicals.

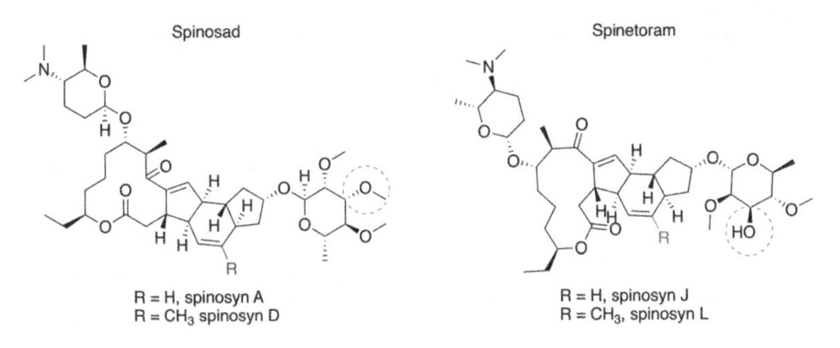


**Figure 8.23** Spinosad and spinetoram.

## Why Is It Superior?

The spinosyns represent an improvement upon existing organophosphate insecticides due to their targeted toxicity, increased potency, and accelerated environmental degradation (Fig. 8.24). As previously discussed, organophosphates are neurotoxins to both insects and mammals. In contrast, the EPA has concluded that spinosad is of low toxicity to mammals and birds.[91] Spinetoram is 1000 or 44 times less acutely toxic to mammals via oral exposure compared to organophosphates azinphos-methyl or phosmet, respectively.[94] Given the high incidence of accidental exposure to organophosphate neurotoxins, spinosad and spinetoram's negative neurotoxicity tests at the highest tested doses[95] ensure a safer job for those who create, mix, and apply insecticides. These spinosyns are also generally much more selective in their toxic effects towards insects, demonstrating a reduced risk to beneficial insects compared to older active ingredients.[93a] Spinosad is reported to leave 70–90% of beneficial insects unharmed.[91] As would be expected, spinosad and spinetoram are still toxic to some non-target species. Between these two pesticides, scientists have found moderate toxicity to fish, ants, earwigs, members of *Hymenoptera* (e.g., bees, wasps), beneficial arthropods, and susceptibility to long-term effects for aquatic invertebrates.[91–93]

**Figure 8.24** Organisms affected by spinosad or spinetoram (mosquito larvae,[96] earwigs,[97] tobacco budworm[98]).

In addition to lower toxicity to non-target species, spinetoram is also effective at a lower concentration than the aforementioned organophosphates. The effective use rates for spinetoram are 10–34 times lower than those for azinphos-methyl and phosmet.[94] This contributed to the projected elimination of 1.8 million pounds of organophosphate insecticides used across the first five years following spinetoram's introduction.[94] Thus, less active pesticide remains on crops. The spinosyns' environmental fate is also generally favorable. Spinosad and spinetoram can photodegrade in water in less than 1 day, and their aerobic half-lives in soil range from a few days to weeks.[95] Due to their low vapor pressure, neither substance is likely to volatilize, and both substances readily adsorb to the soil, reducing their bioavailability.[91,95b] Despite this, the EPA has noted concerns that the toxicity of certain degradants may be more toxic to aquatic organisms.[95a] Taken together, the reduced toxicity to non-target species, greater potency towards target species, and favorable environmental fate make these spinosyns a marked improvement over organophosphate insecticides.

## How, From the Molecular Design Perspective, Was This Accomplished?

The design of the spinosad and spinetoram appears to have focused on the top of the molecular design pyramid, but the selection of the spinosyn molecular scaffold may have reduced absorption and increased excretion. Although the rational design of key events or receptor binding to cause toxicity to one species, but not others, maybe the most challenging aspect of the design pyramid, nature in this case provided a molecular template. Evidently, the search for active insecticide agents, and not minimally toxic agents, produced the first spinosyns. But, from the initial discovery of these macrocyclic lactones, the toxicological tests conducted allowed chemists to exploit the differential toxicity to target and non-target species. Serendipitously, chemists discovered that the novel mechanism for disrupting neurotransmission did not similarly affect many non-target species. Using artificial neural networks to help interpret structure-activity relationships was a further inventive extension of this molecular scaffold.

While Corteva™'s chemist may have focused on bioactivity, the structural backbone that they pursued had some favorable ADME and degradation properties as well. With molecular weights over 700 daltons and log P values between 4.0 and 4.5,[95] per Section 7.5 we may expect that these compounds will not be easily absorbed by the GI or lungs. The low vapor pressures of this substance also ensure minimal exposure via inhalation. As noted in Section 7.15, the labile lactone ester may increase detoxification, and per Section 7.10.1, the presence of many oxygen-containing functional groups may aid in aerobic degradation. Reinforcing this, EPA testing revealed that rat urinal and fecal excretions were nearly completed 48 hours after dosing for spinosad.[95b] While the exact mechanism of action for spinosyns may not be clear, it is also worth noting that the lactone rings in these compounds meet each of the rules given in Fig. 7.30, suggesting that the low toxicities result from these structural characteristics. Regardless of the intentionality in creating the spinosyns, understanding how a unique mode of action, as well as favorable excretion and degradation characteristics, create a better insecticide helps us appreciate future designs for reduced hazard.

## Example IX: Greener Polyurethanes

### What Came Before?

Polyurethanes (PU) are an important class of polymers that have a wide range of applications in household and industrial settings due to their good mechanical properties, low density, and thermal conductivity.[99] They are mainly used as foams, coatings, elastomers, and adhesive sealings.[99a,b] Conventionally, PU are synthesized in the reaction of polyol and polyisocyanate in the presence of a catalyst. However, as we mentioned earlier, isocyanates are very reactive electrophiles that are known as strong sensitizers and cause asthma,[100] and in some cases are suspected carcinogens.[101] Additionally, isocyanates form corrosive and toxic fumes containing hydrogen cyanide and nitrogen oxides when burned.[102] Some of the most important industrial isocyanates are shown in Fig. 8.25, with common applications shown in Fig. 8.26.

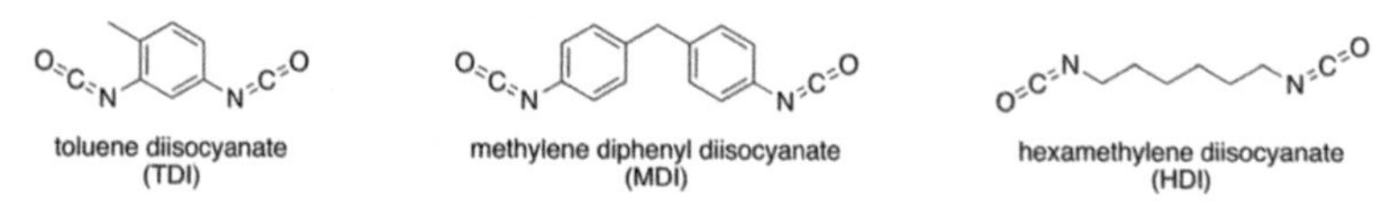


**Figure 8.25** Most common industrial isocyanates.

**Figure 8.26** Polyurethane foam material and paint.[103]

### What Is This New Thing?

Since the production of polyurethane includes toxic isocyanates, one of the significant challenges green chemistry encountered was to develop a process that would be both economically viable and environmentally friendly. In 2015, Hybrid Coating Technologies (HCT) received the Presidential Green Chemistry Challenge Award for developing "Green Polyurethane™" a hybrid non-isocyanate polyurethane (HNIPU) that bears similarities to both conventional polyurethanes and epoxides.[104] HNIPU is formed in the reaction of aliphatic amines and oligomeric cyclocarbonates (Fig. 8.27), and has potential application in foams, composite materials, coatings, paints and varnishes, and adhesives.[105] HCT's procedure involves a reaction between a mixture of epoxy oligomers and mono- or polycyclic carbonates with aliphatic or cycloaliphatic polyamines that have primary amino groups. The result of this reaction is a polymer crosslinked with $\beta$-hydroxyurethane groups. The process of forming HNIPU from diamine and five-membered cyclic carbonates is usually slow at ambient conditions and requires heating above 80 °C. To enhance the formation of a polymer with desired $M_w$, a six-membered ring reaction with diamines using a catalyst, and including side reactions, was investigated in detail by Lambeth et al.[106] Other processes to form HNIPU also exist; however, they either involve the formation of the isocyanates in one of the steps (A–B type azide condensation) or formation of carbamate with polycondensation requiring high temperatures (transurethanization).[99b]

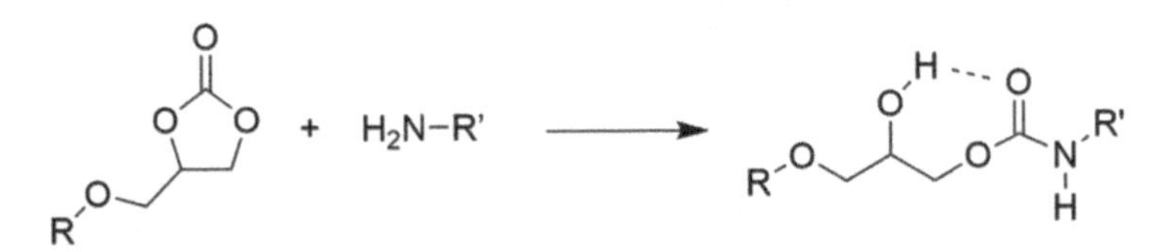


**Figure 8.27** HNIPU synthesis reaction.

### Why Is It Superior?

Some of the main failings of the aminolysis reaction are low $M_W$ polymers and slow kinetics, mostly due to side reactions.[99a] One of the ways to increase the polymer's $M_W$ is by crosslinking polyhydroxyurethanes to form foams as demonstrated by HCT.

While numerous processes to produce isocyanates from renewable sources exist,[99a,d] the non-isocyanate route is much safer and environmentally friendly. In the process of producing HNIPU, HCT employed new hydroxyalkyl urethane modifiers based on renewable vegetable oils, that enable "cold" curing of the epoxy-amine composition. Since it does not include isocyanates, this process is much safer while at the same time, HNIPU has increased chemical and weathering resistance. Additionally, this method provides superior performance characteristics compared to polymer-based acrylic coatings, including shorter drying times, better appearance, and strength-stress properties. 50% of the epoxy base used in the HCT process is sourced from renewables which also makes the HNIPU commercially competitive with conventional epoxy and PU products.[104,107]

### How, From the Molecular Design Perspective, Was This Accomplished?

As we mentioned earlier in the book (Sections 6.3 and 7.18.3), isocyanates are very reactive electrophiles that are known sensitizers and potential carcinogens and they should be avoided. That is why the processes for the production of polyurethanes that do not include isocyanates, both as reagents and also as intermediates, comply with the suggestions for safer chemical design. Exposure to isocyanates is often a consequence of inhalation,[100–102] which is avoided in the case of HNIPU preparation. Generally, as we mentioned in Section 3.10, the presence of heteroatoms and ester bonds in the structure of polymers enable easier degradation, making polyurethanes more environmentally friendly than PET. In the case of HNIPU, the resistance to hydrolysis of the whole urethane group is increased due to the formation of intramolecular hydrogen bonds between the carbonyl oxygen and hydroxyl group on the $\beta$-carbon atom of the PU chain. On the other hand, this structural feature contributes to the chemical resistance of the material.[105b,107]

## Example X: Safer Paints and Coatings

### What Came Before?

Different types of coatings are present everywhere, providing protection to applied surfaces from weathering, corrosion, physical damage, effects of heat, or being used as paints for decorative purposes.[108] A large number of coatings used around the world are made up of polymers based on acrylate monomers or are based on the use of alkyd resin (Figure 8.28, top). Conventional oil-based alkyd paints are high-performing and cost-effective coatings but are prone to yellowing over time—more importantly, they have high VOC content which contributes to significant air pollution due to the formation of ground-level ozone and smog.[108b,c] A large amount of VOC solvent is necessary for solubilizing organic components and getting adequate viscosity for application. Alternative low-VOC alkyd coatings and no VOC solvents exist, but they all suffer from their share of trade-offs, such as inferior performance, increased cost, and undesired odor.[108b] Most of the acrylate that is used for the production of acrylic coatings is produced from fossil-based hydrocarbons in a multi-step process which puts a huge burden on the environment (Figure 8.28, bottom).[108a,109] Additionally, acrylates contain an *α,β*- unsaturated carbonyl system that can undergo Michael addition reaction as we mentioned in Section 7.18.10, which is considered as a cause for the carcinogenicity of acrylates.[110]

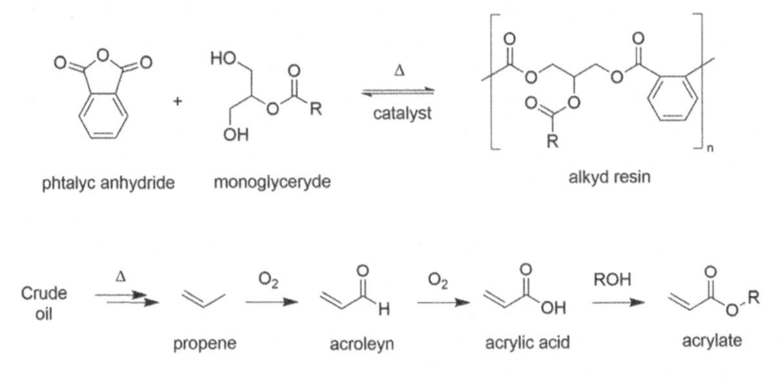


**Figure 8.28** Typical alkyd resin synthesis (top); classic production of acrylates (bottom).

## What Is This New Thing?

In 2009, the Procter & Gamble (PG) and Cook Composites and Polymers (CCP) companies were winners of the Presidential Green Chemistry Challenge for developing innovative Chempol® MPS paint formulations based on the new safer alkyd resin technology that requires less than half of the solvent.[108b] These new formulations are enabled by Sefose® oils that replace petroleum-based solvents by using a renewable feedstock where sucrose is esterified with fatty acids in a solventless process.[108b] The selection of natural oil feedstocks that have an optimal degree of esterification and fatty acid chain length distribution controls the functional density and molecular composition of Sefose® sucrose esters. Sefose® becomes an integral part of the coating film by undergoing auto-oxidative crosslinking with other constituents of the formulation.[108b]

More recently, the Feringa group from the University of Groningen, in collaboration with the Dutch company AkzoNobel, developed a safer and more environmentally friendly process for the production of acrylate coatings from renewable sources.[108a] They used lignocellulose, known to be a good starting material to produce furfural using acid-mediated dehydration,[111] which was then transformed into hydroxybutenolide by photooxidation.[108a] These molecules can easily be transformed into various alkoxybutenolides which bear structural similarity to acrylates and can be (co) polymerized into resins instead of acrylates (Fig. 8.29). The use of different alcohols offers an opportunity to form coatings that will have tunable material properties.[108a]

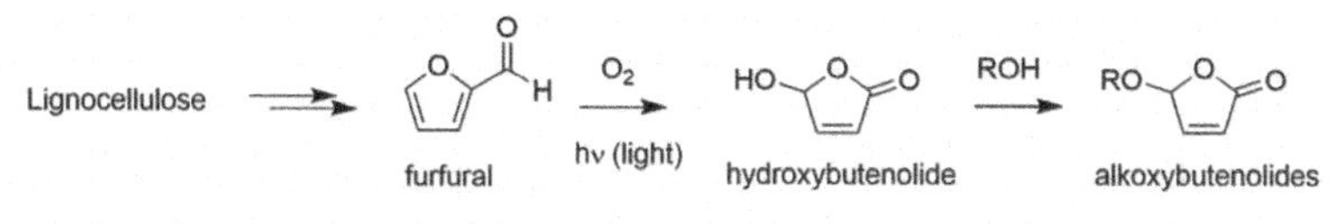


**Figure 8.29** Synthesis of alkoxybutenolide.[108a]

## Why Is It Superior?

With Chempol® MPS alkyd resins, coating formulations demonstrate several advantages such as high gloss, the toughness of film, fast drying times, and more importantly increased renewable content. Due to the significantly lower requirements for solvent thinners, global VOC release is also reduced, which makes new alkyd resins environmentally more friendly than classical alkyd resins. Because Sefose® has a high concentration of fatty acids, incorporating it into coatings contribute to good adhesion and good water resistance. However, the fact remains that solvents are still necessary for the coating applications, which leaves room for finding even better alternatives. This has led to efforts to find new coating systems based on alkyds including nanocomposites, organic-inorganic, and water-borne coatings that demonstrate some improvement compared to traditional alkyd coatings.[112]

The process developed by the Feringa group is an excellent example of a sustainable production route for novel acrylic coatings. It demonstrates the application of several Principles of Green Chemistry by using a less hazardous chemical synthesis starting from renewable materials, catalyzed by light in presence of molecular oxygen in a much safer, and energy-efficient flow reactor.[108a,113] They have shown that alkoxybutenolide monomers synthesized in the process can be used as safer and environmentally friendly alternatives for common acrylates. Additionally, the coatings formed by polymerizing alkoxybutenolides offer tunability for application on different surfaces such as glass and plastics, while demonstrating excellent solvent resistance and hardness. This process is designed in a way that is scalable to industrial levels, offering a path to the commercialization of high-quality coatings that would significantly reduce environmental footprints.

### How, From the Molecular Design Perspective, Was This Accomplished?

As we mentioned in Section 7.6, volatile molecules are easily inhaled, thus exposing organisms to their potentially toxic effects, and a number of VOCs are known environmental pollutants as well.[114] Limiting, or otherwise completely removing, VOCs from coating formulation would be preferable. The new alkyd resins from PG and CCP companies require much less VOCs than classic alkyd resin formulations. This was accomplished by using Sefose® sucrose esters that, due to the possibility of using optimal fatty acids for esterification, achieves desired viscosity, which in turn reduces the need for organic solvents for applications.

Alkoxybutenolide monomers are structurally similar to acrylate, however, they are less reactive which is demonstrated by slower kinetics.[108a] This is a consequence of the double bond being in the internal part of the ring, and the bulky alcohols are present in the structure (Fig. 8.27). These structural features are closely related to the design suggestions we made in Section 7.18.10 to reduce the ability of the $\alpha,\beta$-unsaturated carbonyl group to act as a Michael acceptor. The presence of various alcohol groups made these coatings' properties finely tunable for different purposes. The combination of different monomers enabled the coating of glass or plastic surfaces, and the addition of more rigid monomers yielded harder coatings. From our perspective, one of the most important aspects presented by the work of the Feringa group is the thoughtful molecular design practices and compliance with Principles of Green Chemistry we support throughout this book. By using furfural obtained from lignocellulose, and performing the reaction using benign solvents and light as a catalyst at ambient conditions, they have performed a catalyzed, energy-efficient, and less hazardous synthesis using renewable feedstocks. Additionally, facile derivatization to safe and useful alkoxybutenolides was achieved without the use of activation or protection groups.

## References

1. European Commission. Commission Regulation (EU) 2017/227. *OJEU*, 2017, p. L35/36.
2. Xu, F.; Zhong, L.; Xu, Y.; Zhang, C.; et al. Highly efficient flame-retardant and soft cotton fabric prepared by a novel reactive flame retardant. *Cellulose*, 2019, **26**(6), 4225–4240.
3. van der Veen, I.; de Boer, J. Phosphorus flame retardants: properties, production, environmental occurrence, toxicity and analysis. *Chemosphere*, 2012, **88**(10), 1119–1153.
4. National Research Council. *A Framework to Guide Selection of Chemical Alternatives*. The National Academies Press: Washington, DC, 2014.
5. Technologies, P. Technical Articles. https://www.pyrocooltech.com/pdf/TDFull-rev1.pdf (accessed February 1).
6. Ross, I. Is the burst of the AFFF bubble a precursor to long term environmental liabilities? *International Airport Review* [Online], 2019. Published online: 29 July 2019. https://www.internationalairportreview.com/article/98795/fire-fighting-foam-chemicals-water/ (accessed 2/12/2021).
7. (a) Guo, Y.; Chang, C.-C.; Halada, G.; Cuiffo, M. A.; et al. Engineering flame retardant biodegradable polymer nanocomposites and their application in 3D printing. *Polym Degrad Stab*, 2017, **137**, 205–215; (b) Kolibaba, T. J.; Shih, C.-C.; Lazar, S.; Tai, B. L.; et al. Self-extinguishing

additive manufacturing filament from a unique combination of polylactic acid and a polyelectrolyte complex. *ACS Mater Lett*, 2019, **2**(1), 15–19.

8. (a) Guin, T.; Krecker, M.; Milhorn, A.; Hagen, D. A.; et al. Exceptional flame resistance and gas barrier with thick multilayer nanobrick wall thin films. *Adv Mater Interfaces*, 2015, **2**(11), 1500214; (b) Leistner, M.; Abu-Odeh, A. A.; Rohmer, S. C.; Grunlan, J. C. Water-based chitosan/melamine polyphosphate multilayer nanocoating that extinguishes fire on polyester-cotton fabric. *Carbohydr Polym*, 2015, **130**, 227–232; (c) Holder, K. M.; Cain, A. A.; Plummer, M. G.; Stevens, B. E.; et al. Carbon nanotube multilayer nanocoatings prevent flame spread on flexible polyurethane foam. *Macromol Mater Eng*, 2016, **301**(6), 665–673; (d) Lazar, S.; Carosio, F.; Davesne, A. L.; Jimenez, M.; et al. Extreme heat shielding of clay/chitosan nanobrick wall on flexible foam. *ACS Appl Mater Interfaces*, 2018, **10**(37), 31686–31696; (e) Liu, X.; Qin, S.; Li, H.; Sun, J.; et al. Combination intumescent and kaolin-filled multilayer nanocoatings that reduce polyurethane flammability. *Macromol Mater Eng*, 2018, **304**(2), 1800531.
9. (a) Haile, M.; Fincher, C.; Fomete, S.; Grunlan, J. C. Water-soluble polyelectrolyte complexes that extinguish fire on cotton fabric when deposited as pH-cured nanocoating. *Polym Degrad Stab*, 2015, **114**, 60–64; (b) Leistner, M.; Haile, M.; Rohmer, S.; Abu-Odeh, A.; et al. Water-soluble polyelectrolyte complex nanocoating for flame retardant nylon-cotton fabric. *Polym Degrad Stab*, 2015, **122**, 1–7; (c) Haile, M.; Leistner, M.; Sarwar, O.; Toler, C. M.; et al. A wash-durable polyelectrolyte complex that extinguishes flames on polyester–cotton fabric. *RSC Adv*, 2016, **6**(40), 33998–34004.
10. (a) Malucelli, G. Biomacromolecules and bio-sourced products for the design of flame retarded fabrics: current state of the art and future perspectives. *Molecules*, 2019, **24**(20), 3744; (b) Riehle, F.; Hoenders, D.; Guo, J.; Eckert, A.; et al. Sustainable chitin nanofibrils provide outstanding flame-retardant nanopapers. *Biomacromolecules*, 2019, **20**(2), 1098–1108.
11. Khandual, A. Green flame retardants for textiles. In: Muthu, S. G., M. (Ed.), *Green Fashion. Environmental Footprints and Eco-design of Products and Processes*, Springer: Singapore, 2016, pp. 171–227.
12. Rizos, C. V.; Elisaf, M. S.; Mikhailidis, D. P.; Liberopoulos, E. N. How safe is the use of thiazolidinediones in clinical practice? *Expert Opin Drug Saf* 2009, **8**(1), 15–32.
13. Hasan, F. M.; Alsahli, M.; Gerich, J. E. SGLT2 inhibitors in the treatment of type 2 diabetes. *Diabetes Res Clin Pract*, 2014, **104**(3), 297–322.

14. (a) Diamant, M.; Morsink, L. M. SGLT2 inhibitors for diabetes: turning symptoms into therapy. *Lancet*, 2013, **382**(9896), 917–918; (b) Bauer, A.; Bronstrup, M. Industrial natural product chemistry for drug discovery and development. *Nat Prod Rep*, 2014, **31**(1), 35–60.

15. Wilding, J. P. The role of the kidneys in glucose homeostasis in type 2 diabetes: clinical implications and therapeutic significance through sodium glucose co-transporter 2 inhibitors. *Metabolism*, 2014, **63**(10), 1228–1237.

16. Food and Drug Administration. FDA revises labels of SGLT2 inhibitors for diabetes to include warnings about too much acid in the blood and serious urinary tract infections. https://www.fda.gov/drugs/drug-safety-and-availability/fda-revises-labels-sglt2-inhibitors-diabetes-include-warnings-about-too-much-acid-blood-and-serious (accessed February 1).

17. (a) Washburn, W. N. Development of the renal glucose reabsorption inhibitors: a new mechanism for the pharmacotherapy of diabetes mellitus type 2. *J Med Chem*, 2009, **52**(7), 1785–94; (b) Blaschek, W. Natural products as lead compounds for sodium glucose cotransporter (SGLT) inhibitors. *Planta Med*, 2017, **83**(12–13), 985–993.

18. Meng, W.; Ellsworth, B. A.; Nirschl, A. A.; McCann, P. J.; et al. Discovery of dapagliflozin: a potent, selective renal sodium-dependent glucose cotransporter 2 (SGLT2) inhibitor for the treatment of type 2 diabetes. *J Med Chem*, 2008, **51**(5), 1145–1149.

19. (a) Hongu, M.; Funami, N.; Takahashi, Y.; Saito, K.; et al. Na(+)-glucose cotransporter inhibitors as antidiabetic agents. III. Synthesis and pharmacological properties of 4′-dehydroxyphlorizin derivatives modified at the OH groups of the glucose moiety. *Chem Pharm Bull (Tokyo)*, 1998, **46**(10), 1545–1555; (b) Tsujihara, K.; Hongu, M.; Saito, K.; Kawanishi, H.; et al. Na(+)-glucose cotransporter (SGLT) inhibitors as antidiabetic agents. 4. Synthesis and pharmacological properties of 4′-dehydroxyphlorizin derivatives substituted on the B ring. *J Med Chem*, 1999, **42**(26), 5311–5324.

20. Katsuno, K.; Fujimori, Y.; Takemura, Y.; Hiratochi, M.; et al. Sergliflozin, a novel selective inhibitor of low-affinity sodium glucose cotransporter (SGLT2), validates the critical role of SGLT2 in renal glucose reabsorption and modulates plasma glucose level. *J Pharmacol Exp Ther*, 2007, **320**(1), 323–330.

21. Fujimori, Y.; Katsuno, K.; Nakashima, I.; Ishikawa-Takemura, Y.; et al. Remogliflozin etabonate, in a novel category of selective low-affinity sodium glucose cotransporter (SGLT2) inhibitors, exhibits antidiabetic

efficacy in rodent models. *J Pharmacol Exp Ther*, 2008, **327**(1), 268–276.

22. da Silva, P. N.; da Conceicao, R. A.; do Couto Maia, R.; de Castro Barbosa, M. L. Sodium-glucose cotransporter 2 (SGLT-2) inhibitors: a new antidiabetic drug class. *Medchemcomm*, 2018, **9**(8), 1273–1281.
23. Mori, T.; Ito, T.; Liu, S.; Ando, H.; et al. Structural basis of thalidomide enantiomer binding to cereblon. *Sci Rep*, 2018, **8**(1), 1294.
24. Celgene Corporation. Safety. https://revlimidhcp.com/mds/safety (accessed February 1).
25. Kotla, V.; Goel, S.; Nischal, S.; Heuck, C.; et al. Mechanism of action of lenalidomide in hematological malignancies. *J Hematol Oncol*, 2009, **2**, 36.
26. Holstein, S. A.; McCarthy, P. L. Immunomodulatory drugs in multiple myeloma: mechanisms of action and clinical experience. *Drugs*, 2017, **77**(5), 505–520.
27. (a) Latif, T.; Chauhan, N.; Khan, R.; Moran, A.; et al. Thalidomide and its analogues in the treatment of multiple myeloma. *Exp Hematol Oncol*, 2012, **1**(1), 27; (b) Chang, X.; Zhu, Y.; Shi, C.; Stewart, A. K. Mechanism of immunomodulatory drugs' action in the treatment of multiple myeloma. *Acta Biochim Biophys Sin (Shanghai)*, 2014, **46**(3), 240–253; (c) Sherbet, G. V. Therapeutic potential of thalidomide and its analogues in the treatment of cancer. *Anticancer Res*, 2015, *35*, 5767–5772; (d) Drahy, F.; Ingen-Housz-Oro, S.; Grootenboer-Mignot, S.; Wolkenstein, P.; et al. Lenalidomide as an alternative to thalidomide for treatment of recurrent erythema multiforme. *JAMA Dermatol*, 2018, **154**(4), 487–489.
28. Smith, R. L.; Mitchell, S. C. Thalidomide-type teratogenicity: structure-activity relationships for congeners. *Toxicol Res (Camb)*, 2018, **7**(6), 1036–1047.
29. Schultz, T. P.; Nicholas, D. D.; Preston, A. F. A brief review of the past, present and future of wood preservation. *Pest Manag Sci*, 2007, **63**(8), 784–788.
30. United States Environmental Protection Agency. Overview of wood preservative chemicals. https://www.epa.gov/ingredients-used-pesticide-products/overview-wood-preservative-chemicals (accessed January 25).
31. Schultz, T. P.; Nicholas, D. D. Development of environmentally-benign wood preservatives based on the combination of organic biocides with antioxidants and metal chelators. *Phytochemistry*, 2002, **61**(5), 555–560.

32. United States Environmental Protection Agency. ACQ Preserve®: the environmentally advanced wood preservative. https://www.epa.gov/greenchemistry/presidential-green-chemistry-challenge-2002-designing-greener-chemicals-award (accessed January 25).

33. Mohajerani, A.; Vajna, J.; Ellcock, R. Chromated copper arsenate timber: a review of products, leachate studies and recycling. *J Clean Prod*, 2018, **179**, 292–307.

34. Song, J.; Dubey, B.; Jang, Y.-C.; Townsend, T.; et al. Implication of chromium speciation on disposal of discarded CCA-treated wood. *J Hazard Mater*, 2006, **128**(2), 280–288.

35. Hiker, G. S. Rotted dock (CC BY 2.0). https://www.flickr.com/photos/42693172@N05/7321156266 (accessed January 25).

36. Schmitt, S.; Zhang, J.; Shields, S.; Schultz, T. Chapter 12: Copper-based wood preservative systems used for residential applications in North America and Europe. *Deterioration and Protection of Sustainable Biomaterials*, ACS Symposium Series, Vol. 1158, American Chemical Society, 2014, pp. 217–225.

37. Hora, P. I.; Pati, S. G.; McNamara, P. J.; Arnold, W. A. Increased use of quaternary ammonium compounds during the SARS-CoV-2 pandemic and beyond: consideration of environmental implications. *Environ Sci Technol Lett*, 2020, **7**(9), 622–631.

38. Freeman, M.; Nicholas, D.; Schultz, T. Non-arsenical wood protection: alternatives for CCA, creosote, and pentachlorophenol. In: Townsend, T. G.; Solo-Gabriele, H. (Eds.), *Envrionmental Impacts of Treated Wood*, Taylor & Francis: Boca Raton, FL, 2006.

39. Goodell, B. Chapter 6: Brown-rot fungal degradation of wood: our evolving view. *Wood Deterioration and Preservation*, ACS Symposium Series, Vol. 845, American Chemical Society, 2003, pp. 97–118.

40. Tullo, A. H. Making wood last forever with acetylation. *c&en Biomaterials* [Online], 2012, **90**. https://cen.acs.org/articles/90/i32/Making-Wood-Last-Forever-Acetylation.html.

41. Mantanis, G. Chemical modification of wood by acetylation or furfurylation: a review of the present scaled-up technologies. *BioResources*, 2017, **12**(2), 4478–4489.

42. Luz, A.; DeLeo, P.; Pechacek, N.; Freemantle, M. Human health hazard assessment of quaternary ammonium compounds: didecyl dimethyl ammonium chloride and alkyl (C12–C16) dimethyl benzyl ammonium chloride. *Regul Toxicol Pharmacol*, 2020, **116**, 104717.

43. (a) PubChem. Benzyldimethyltetradecylammonium chloride. National Center for Biotechnology Information: Online, 2021; (b) PubChem. Didecyldimethylammonium chloride. National Center for Biotechnology Information: Online, 2021.
44. United States Environmental Protection Agency. Designing an environmentally safe marine antifoulant. https://www.epa.gov/greenchemistry/presidential-green-chemistry-challenge-1996-designing-greener-chemicals-award (accessed February 1).
45. Abarzua, S.; Jakubowski, S. Biotechnological investigation for the prevention of biofouling. I. Biological and biochemical principles for the prevention of biofouling. *Mar Ecol Prog Ser*, 1995, **123**, 301–312.
46. Callow, M. E.; Willingham, G. L. Degradation of antifouling biocides. *Biofouling*, 1996, **10**(1–3), 239–249.
47. Laughlin, R. B.; Guard, H. E.; Coleman, W. M. Tributyltin in seawater: speciation and octanol-water partition coefficient. *Environ Sci Technol*, 1986, **20**(2), 201–204.
48. Holleman, J. Research to application: oyster shell research leads to patent to deter biofouling. https://seagrant.noaa.gov/News/Article/ArtMID/1660/ArticleID/626/Research-to-Application-Oyster-shell-research-leads-to-patent-to-deter-biofouling (accessed Feb 11).
49. Willingham, G. L.; Jacobson, A. H. Chapter 11: Designing an environmentally safe marine antifoulant. *Designing Safer Chemicals*, ACS Symposium Series, Vol. 640, American Chemical Society, 1996, pp. 224–233.
50. Silva, V.; Silva, C.; Soares, P.; Garrido, E. M.; et al. Isothiazolinone biocides: chemistry, biological, and toxicity profiles. *Molecules*, 2020, **25**(4), 991.
51. Jacobson, A. H.; Willingham, G. L. Sea-nine antifoulant: an environmentally acceptable alternative to organotin antifoulants. *Sci Total Environ*, 2000, **258**(1), 103–110.
52. Gabe, H. B.; Guerreiro, A. d. S.; Sandrini, J. Z. Molecular and biochemical effects of the antifouling DCOIT in the mussel *Perna perna*. *Comp Biochem Physiol C Toxicol Pharmacol*, 2021, **239**, 108870.
53. European Chemicals Agency. 4,5-dichloro-2-octyl-2H-isothiazol-3-one. https://echa.europa.eu/substance-information/-/substanceinfo/100.058.930 (accessed February 3).
54. Chen, L.; Lam, J. C. W. SeaNine 211 as antifouling biocide: a coastal pollutant of emerging concern. *J Environ Sci*, 2017, **61**, 68–79.

55. Shade, W.; Hurt, S.; Jacobson, A.; Reinert, K. Ecological risk assessment of a novel marine antifoulant. In: Gorsuch, J. W.; Dwyer, F. J.; Ingersoll, C. G.; Point, T. W. L. (Eds.), *Environmental Toxicology and Risk Assessment*, Vol. 2, ASTM STP 1216, American Society for Testing and Materials: Philidelphia, 1993.

56. Chen, L.; Ye, R.; Xu, Y.; Gao, Z.; et al. Comparative safety of the antifouling compound butenolide and 4,5-dichloro-2-n-octyl-4-isothiazolin-3-one (DCOIT) to the marine medaka (*Oryzias melastigma*). *Aquat Toxicol*, 2014, **149**, 116–125.

57. Eaton, D. E.; Clesceri, L. S.; Greenberg, A. E.; Franson, M. A. H.; et al. *Standard methods for the examination of water and wastewater*. 20th ed., American Public Health Association: Washington, DC, 1998.

58. Palmer, M.; Hatley, H. The role of surfactants in wastewater treatment: impact, removal and future techniques: a critical review. *Water Res*, 2018, **147**, 60–72.

59. Rain, B. "Water Stinks," Writes the Photographer About this Scene. National Archives at College Park: United States National Archives Catalog, 1972.

60. Boethling, R. S.; Sommer, E.; DiFiore, D. Designing small molecules for biodegradability. *Chem Rev*, 2007, **107**(6), 2207–2227.

61. Rosen, M. J.; Li, F.; Morrall, S. W.; Versteeg, D. J. The relationship between the interfacial properties of surfactants and their toxicity to aquatic organisms. *Environ Sci Technol*, 2001, **35**(5), 954–959.

62. Nielsen, A. D.; Borch, K.; Westh, P. Thermochemistry of the specific binding of C12 surfactants to bovine serum albumin. *Biochim Biophys Acta, Protein Struct Mol Enzymol*, 2000, **1479**(1–2), 321–331.

63. Bhattacharya, S.; Mandal, S. S. Interaction of surfactants with DNA. Role of hydrophobicity and surface charge on intercalation and DNA melting. *Biochim Biophys Acta, Biomembr*, 1997, **1323**(1), 29–44.

64. Shim, Y. J.; Choi, J. H.; Ahn, H. J.; Kwon, J. S. Effect of sodium lauryl sulfate on recurrent aphthous stomatitis: a randomized controlled clinical trial. *Oral Dis*, 2012, **18**(7), 655–660.

65. Effendy, I.; Maibach, H. I. Surfactants and experimental irritant contact dermatitis. *Contact Derm*, 1995, **33**(4), 217–225.

66. United States Environmental Protection Agency. DfE alternatives assessment for nonylphenol ethoxylates. U. S. EPA, May 2012.

67. Voutchkova, A. M.; Osimitz, T. G.; Anastas, P. T. Toward a comprehensive molecular design framework for reduced hazard. *Chem Rev*, 2010, **110**(10), 5845–5882.

68. United States Environmental Protection Agency. EPA's safer choice standard, revision 2015 ed., U. S. EPA, 2015, p. 41.

69. Rebello, S.; Asok, A. K.; Mundayoor, S.; Jisha, M. S. Surfactants: toxicity, remediation and green surfactants. *Environ Chem Lett*, 2014, **12**(2), 275–287.

70. (a) Li, G.; Chen, L.; Ruan, Y.; Guo, Q. Alkyl polyglycoside: a green and efficient surfactant for enhancing heavy oil recovery at high-temperature and high-salinity condition. *J Pet Explor Prod Technol*, 2019, **9**(4), 2671–2680; (b) Geetha, D.; Tyagi, R. Alkyl poly glucosides (APGs) surfactants and their properties: a review. *Tenside Surfact Det*, 2012, **49**(5), 417–427.

71. DOW Chemical. *TRITON™ CG-425 Alkyl Polyglucoside Surfactant;* 805-00047-0614 CDP, Dow Consumer & Industrial Solutions, Online 2014.

72. Eslami, P.; Hajfarajollah, H.; Bazsefidpar, S. Recent advancements in the production of rhamnolipid biosurfactants by *Pseudomonas aeruginosa*. *RSC Adv*, 2020, **10**(56), 34014–34032.

73. United States Environmental Protection Agency. Rhamnolipid biosurfactant: a natural, low-toxicity alternative to synthetic surfactants. https://www.epa.gov/greenchemistry/2004-small-business-award (accessed February 1).

74. Asok, A. K.; Jisha, M. S. Biodegradation of the anionic surfactant linear alkylbenzene sulfonate (LAS) by autochthonous *Pseudomonas* sp. *Water Air Soil Pollut*, 2012, **223**(8), 5039–5048.

75. (a) Madsen, T.; Petersen, G.; Seierø, C.; Tørsløv, J. Biodegradability and aquatic toxicity of glycoside surfactants and a nonionic alcohol ethoxylate. *JAOCS*, 1996, **73**(7), 929–933; (b) Campbell, P. *Alternatives to Nonylphenol Ethoxylates: Review of Toxicity, Biodegradation & Technical-Economic Aspects*. ToxEcology Environmental Consulting Limited: Canada, 2002.

76. Randhawa, K. K. S.; Rahman, P. K. S. M. Rhamnolipid biosurfactants: past, present, and future scenario of global market. *Front Microbiol*, 2014, **5**, 454.

77. Dixon, N. J. Greener chelating agents. In: Anastas, P. T.; Boethling, R.; Voutchkova, A. (Eds.), *Handbook of Green Chemistry*, Wiley-VCH: Weinheim, Germany, 2012, pp. 281–307.

78. (a) Nowack, B.; VanBriesen, J. M. Chelating agents in the environment. In: Nowack, B.; VanBriesen, J. M. (Eds.), *Biogeochemistry of Chelating Agents*, Vol. 910, American Chemical Society: Washington, DC, 2005, pp. 1–18; (b) Sillanpaa, M. Environmental fate of EDTA and DTPA. *Rev Environ Contam Toxicol*, 1997, **152**, 85–111.

79. Bayer Corporation. Baypure$^{TM}$ CX (sodium iminodisuccinate): an environmentally friendly and readily biodegradable chelating agent. https://www.epa.gov/greenchemistry/presidential-green-chemistry-challenge-2001-greener-synthetic-pathways-award (accessed February).

80. Schowanek, D.; Feijtel, T. C. J.; Perkins, C. M.; Hartman, F. A.; et al. Biodegradation of [S,S], [R,R] and mixed stereoisomers of ethylene diamine disuccinic acid (EDDS), a transition metal chelator. *Chemosphere*, 1997, **34**(11), 2375–2391.

81. (a) United States Environmental Protection Agency. Final designation of low-priority substances under the Toxic Substances Control Act (TSCA), Notice of Availability. Federal Register, Vol. 85, U. S. EPA, 2020, pp. 11069–11079; (b) European Commission. E 576: Sodium gluconate. *Food Addititves, OJEU*, Online, 2019; (c) Asemave, K. Greener chelators for recovery of metals and other applications. *OMCIJ*, 2018, **6**(4), 92–102.

82. (a) Balzano, S.; Sardo, A.; Blasio, M.; Chahine, T. B.; et al. Microalgal metallothioneins and phytochelatins and their potential use in bioremediation. *Front Microbiol*, 2020, **11**, 517; (b) Pal, R.; Rai, J. P. Phytochelatins: peptides involved in heavy metal detoxification. *Appl Biochem Biotechnol*, 2010, **160**(3), 945–963.

83. (a) Cao, A.; Carucci, A.; Lai, T.; Bacchetta, G.; et al. Use of native species and biodegradable chelating agents in the phytoremediation of abandoned mining areas. *J Chem Technol Biotechnol*, 2009, **84**(6), 884–889; (b) Cokesa, Z.; Knackmuss, H. J.; Rieger, P. G. Biodegradation of all stereoisomers of the EDTA substitute iminodisuccinate by *Agrobacterium tumefaciens* BY6 requires an epimerase and a stereoselective C-N lyase. *Appl Environ Microbiol*, 2004, **70**(7), 3941–3947.

84. (a) Kolodynska, D. Application of a new generation of complexing agents in removal of heavy metal ions from different wastes. *Environ Sci Pollut Res Int*, 2013, **20**(9), 5939–5949; (b) Tandy, S.; Bossart, K.; Mueller, R.; Ritschel, J.; et al. Extraction of heavy metals from soils using biodegradable chelating agents. *Environ Sci Technol*, 2004, **38**(3), 937–944; (c) Jones, P. W.; Williams, D. R. Chemical speciation simulation used to assess the efficiency of environment-friendly EDTA alternatives for use in the pulp and paper industry. *Inorg Chim Acta*, 2002, **339**, 41–50.

85. (a) Rabie, A. I.; El-Din, H. A. *Sodium Gluconate as a New Environmentally Friendly Iron Controlling Agent for HP/HT Acidizing Treatments*, SPE Middle East Oil & Gas Show and Conference, Manama, Bahrain, Society

of Petroleum Engineers, Manama, Bahrain, 2015; (b) Ramachandran, S.; Fontanille, P.; Pandey, A.; Larroche, C. Gluconic acid: properties, applications and microbial production. *Food Technol Biotech*, 2006, **44**(2), 185–195.

86. Office of Pollution Prevention and Toxics. Supporting information for low-priority substance D-gluconic acid, sodium salt (1:1) (CASRN 527-07-1) (sodium gluconate). U. S. EPA: Washington, DC, 2020, p. 138.
87. Nurchi, V. M.; Cappai, R.; Crisponi, G.; Sanna, G.; et al. Chelating agents in soil remediation: a new method for a pragmatic choice of the right chelator. *Front Chem*, 2020, **8**, 597400.
88. Alavanja, M. C. R. Introduction: pesticides use and exposure extensive worldwide. *Rev Environ Health*, 2009, **24**(4), 303–309.
89. Costa, L. G. Current issues in organophosphate toxicology. *Clin Chim Acta*, 2006, **366**(1), 1–13.
90. Costa, L. G. Organophosphorus compounds at 80: some old and new issues. *Toxicol Sci*, 2017, **162**(1), 24–35.
91. Agriscience, C. Spinosad: a new natural product for insect control. *Presidential Green Chemistry Challenge*, U. S. EPA, 1999.
92. Biondi, A.; Mommaerts, V.; Smagghe, G.; Viñuela, E.; The non-target impact of spinosyns on beneficial arthropods. *Pest Manag Sci*, 2012, **68**(12), 1523–1536.
93. (a) Jean-Pierre, L.; Faiek, E.; Kevin, F.; Jorg, R.; et al. A review on the toxicity and non-target effects of macrocyclic lactones in terrestrial and aquatic environments. *Curr Pharm Biotechnol*, 2012, **13**(6), 1004–1060; (b) Dayan, F. E.; Cantrell, C. L.; Duke, S. O. Natural products in crop protection. *Bioorg Med Chem*, 2009, **17**(12), 4022–4034.
94. United States Environmental Protection Agency. Spinetoram: Enhancing a Natural Product for Insect Control. https://www.epa.gov/greenchemistry/presidential-green-chemistry-challenge-2008-designing-greener-chemicals-award (accessed February 6).
95. (a) United States Environmental Protection Agency. Pesticide Fact Sheet: Spinetoram, U. S. EPA, Online, 2009, p. 13; (b) United States Environmental Protection Agency. Pesticide Fact Sheet: Spinosad, U. S. EPA, Online, 2000.
96. USGS Bee Inventory and Monitoring Lab. Fresh mosquito larvae (CC0). https://www.flickr.com/photos/54563451@N08/9764581325 (accessed February 15).
97. Bullock, T. Earwig Macro (CC BY 2.0). https://www.flickr.com/photos/tombullock/30753260936/in/photolist-NRyv3Q-2acrxLU-nxgWR3-

q9HBeV-awjdKu-2j2VMkX-5g6dC-aTCUvT-4AwYND-bcwu56-c5BLYo-23ymfzc-NRyuvh-t2vtn-aTCUwa-5Q2TW7-5gT4JR-fk7fgN-kL9GZP-b-pys4h-Ssz85E-2jYbJdf-5PXD8r-RtCkNm-4gH8fH-QhW9px-bCk2Fo-vpcuCT-6XrLjP-6wScd6-6XvMkm-5PXD6k-cPT1GW-cTMmv7-PG6FW1-PJRFy6-gkMSsq-tUn29-gkMuDo-f5GxCK-dv21HG-2j6YA6w-2i5zCpA-24xZkdV-27sLNzn-2i19Jzj-2keU78g-wkpqym-XAf1v8-i5Dw7 (accessed May 6).

98. Cranshaw, W. tobacco budworm (Heliothis virescens) (Fabricius) (CC BY 3.0 U.S.). https://www.insectimages.org/browse/detail.cfm?imgnum=5422147 (accessed February 15).

99. (a) Konieczny, J.; Loos, K. Green polyurethanes from renewable isocyanates and biobased white dextrins. *Polymers (Basel)*, 2019, **11**(2), 256; (b) Furtwengler, P.; Avérous, L. Renewable polyols for advanced polyurethane foams from diverse biomass resources. *Polym Chem*, 2018, **9**(32), 4258–4287; (c) Guillame, S. M.; Khalil, H.; Misra, M. Green and sustainable polyurethanes for advanced applications. *J Appl Polym Sci*, 2017, **134**(45), 45646; (d) Agrawal, A.; Kaur, R.; Walia, R. S. PU foam derived from renewable sources: perspective on properties enhancement— An overview. *Eur Polym J*, 2017, **95**, 255–274.

100. National Institute for Occupational Safety and Health. *Preventing Asthma and Death from MDI* Exposure During Spray-on Truck Bed Liner and Related Applications*, 2006–149; United States Centers for Disease Control and Prevention: Cincinnati, OH, 2006, p. 42.

101. Bolognesi, C.; Baur, X.; Marczynski, B.; Norppa, H.; et al. Carcinogenic risk of toluene diisocyanate and 4,4′-methylenediphenyl diisocyanate: epidemiological and experimental evidence. *Crit Rev Toxicol*, 2001, **31**(6), 737–772.

102. Bengtström, L.; Salden, M.; Stec, A. A. The role of isocyanates in fire toxicity. *Fire Sci Rev*, 2016, **5**(4).

103. (a) Old Photo Profile. Plain blue and pink polyurethane kitchen sponges (CC BY 2.0). https://www.flickr.com/photos/10361931@N06/4273918578 (accessed March 1); (b) Soft Surfaces Ltd. Polyurethane paint polymeric sports surface MUGA (CC BY 2.0). https://www.flickr.com/photos/68055546@N03/6210746839 (accessed March 1).

104. United States Environmental Protection Agency. Hybrid non-isocyanate polyurethane/green polyurethane™. https://www.epa.gov/greenchemistry/presidential-green-chemistry-challenge-2015-designing-greener-chemicals-award (accessed February 1).

105. (a) Lavoie, E.; DiFiore, D.; Marshall, M.; Lin, C.; et al. Informing substitution to safer alternatives. In: Anastas, P. T.; Boethling, R.; Voutchkova, A. (Eds.), *Handbook of Green Chemistry*, Vol. 9, Wiley-VCH: Weinheim, Germany, 2012, pp. 107–136; (b) Figovsky, O. L.; Shapovalov, L.; Axenov, O. Advanced coatings based upon non-isocyanate polyurethanes for industrial applications. *Surf Coat Int B: Coat Trans*, 2004, **87**(2), 83–90.

106. Lambeth, R. H.; Mathew, S. M.; Baranoski, M. H.; Housman, K. J.; et al. Nonisocyanate polyurethanes from six-membered cyclic carbonates: catalysis and side reactions. *J Appl Polym Sci*, 2017, **134**(45), 44941.

107. Nanotech Industries Inc. Green polyurethane™. https://nanotechindustriesinc.com/GPU.php (accessed 02/18/2021).

108. (a) Hermens, J. G. H.; Freese, T.; van den Berg, K. J.; van Gemert, R.; et al. A coating from nature. *Sci Adv* 2020, **6**(51); (b) United States Environmental Protection Agency. Chempol® MPS resins and Sefose® sucrose esters enable high-performance low-VOC alkyd paints and coatings. https://www.epa.gov/greenchemistry/presidential-green-chemistry-challenge-2009-designing-greener-chemicals-award (accessed February 1); (c) Jones, F. N. Alkyd resins. *Ullmann's Encyclopedia of Industrial Chemistry*, Wiley-VCH, Online, 2003; (d) Soucek, M. D.; Salata, R. R. Alkyd resin synthesis. In: Kobayashi, S.; Müllen, K. (Eds.) *Encyclopedia of Polymeric Nanomaterials*, Springer: Berlin, 2014, pp. 1–6.

109. Alfke, G.; Irion, W. W.; Neuwirth, O. S. Oil refining. *Ullmann's Encyclopedia of Industrial Chemistry*, Wiley-VCH, Online, 2007.

110. United States Environmental Protection Agency. Provisional peer-reviewed toxicity values for ethyl acrylate. U. S. EPA: Cincinnati, OH, 2014, p. 89.

111. Zeitsch, K. J. *The Chemistry and Technology of Furfural and Its Many By-Product*, 1 ed., Vol. 13, Elsevier Science, Online, 2000, p. 376.

112. Alam, M.; Akram, D.; Sharmin, E.; Zafar, F.; et al. Vegetable oil based eco-friendly coating materials: a review article. *Arab J Chem*, 2014, **7**(4), 469–479.

113. Sambiagio, C.; Noël, T. Flow photochemistry: shine some light on those tubes! *Trends Chem*, 2020, **2**(2), 92–106.

114. *Volatile Organic Compounds in Environment*. MDPI: 2017.

# Chapter 9

# The Path Forward

This book has laid out a basis for chemists to use in designing chemicals so they can strive toward minimal hazards whether they are physical, toxicological, or global. The framework presented is based on the recognition that it is the combination of properties that a molecule possesses that decides whether it is going to be harmful to people and the planet and, if so, in what ways. This approach recognizes that what chemists have been doing for centuries is controlling the structure in increasingly sophisticated ways in order to control those properties. While this has historically been used to achieve certain performance and functions by our chemicals and materials, it is now being used increasingly to avoid hazards of various types.

There have been tremendous achievements in reducing chemical hazards even if some of these advances have been serendipitous. Consider the fact that over the course of the past century we have moved away from:

- Chronic mercury poisoning in the hatmaker's profession causing neurotoxicity and the phrase "Mad as a Hatter"[1]
- Deadly dry-cleaning where the solvents used most prevalently were either toxic, flammable, or both[2]
- Sinister Scheele's green—a poisonous color[3]
- Building insulation with dangerous asbestos[4]

*First Do No Harm: A Chemist's Guide to Molecular Design for Reduced Hazard*
Predrag V. Petrovic and Paul T. Anastas

ISBN 978-981-4968-59-1 (Hardcover), 978-1-003-35964-7 (eBook)
www.jennystanford.com

- Radioactive toothpastes, laxatives, and wrist watches[5]
- Lead used so ubiquitously and dispersed so insidiously that the damage to the brains of the exposed youths is impossible to quantify[6]

The previous chapter of this book showed how we can move from serendipity to design and how we can reduce hazards in everything ranging from polymers to pesticides and everything in between.

**However, it is important to recognize that although molecular design is a nascent field in its infancy, the power and potential for this field to grow and bring about positive benefits in the future is immense. It is equally important to recognize that because it is in its infancy, the tools and methods of design still need to be verified with empirical testing prior to any release of a substance to the environment.**

The scientific and implementation challenges that need to be addressed to advance the field of molecular design for reduced hazards are numerous and are, in some cases, daunting.[7] These challenges include but are not limited to:

- Mechanism—For every structural class or subclass of chemicals that we have well-elucidated mechanisms of action (MOA) or adverse outcome pathways (AOPs), there are thousands for which we have little or no insight. These mechanistic insights at the molecular level are what give chemists the knowledge to have the most degrees of freedom to design structure, and hence properties, with the greatest confidence of reduced hazard. There needs to be extensive continued investment and expansion of this research field to empower molecular design for reduced hazards.
- Mixtures—While it is crucial to understand mechanisms of hazard for individual chemicals, it is also critical to understand the consequences of chemicals in combination. Real-world scenarios almost never involve exposure to a single chemical but rather complex combinations of chemicals in the air we breathe, the food we eat, the pharmaceuticals we consume, etc. There is significant knowledge that substances interact and that they can interact in different ways. The knowledge of when substances will magnify or offset various adverse effects when in mixtures is crucial to real-world benefits.

- Vulnerability—There is a deepening understanding that every organism is not equally vulnerable to all hazards at all stages of life. A more rigorous understanding of these windows of vulnerability is crucial to future design.
- Non-standard dose-response—Substances that have shown endocrine-disrupting effects are often also demonstrating dose-response curves that question traditional paradigms. We need an investigation into effects at low doses that may counter-intuitively not appear at higher doses.
- Models—Our biological models informing our computational models can hold tremendous promise for the future but require significant development. Even our most advanced computer modelling possesses limitations as fundamental as the ability to account for many of the most important biological intermolecular interactions.
- Systems thinking—Perhaps one of the greatest scientific challenges is moving from the belief that we can understand problems using reductionist methodologies and that they can be directly applied to systems problems and systems design. In the immense complexity of biological systems of the biosphere interacting with the physical systems of the geosphere, there needs to be a translation of our models and assays into the broader context. Failure to confront this limitation with humility will likely result in potentially serious unintended consequences.

While these enumerated scientific challenges and others need to be addressed as the field of molecular design evolves, there are other practical challenges for the full adoption of this science to reach its potential positive impact. It is important that the scientific achievements in this area are never limited to merely academic insights whose influence do not leave the scientific literature.

Perhaps the most important step in achieving this is that there needs to be a general awareness that safer chemicals are *possible.* Today, there exists a general belief—a powerful myth—that the substances used in society are there because it is necessary to use them. If there is a toxic substance in a product or a manufacturing process, it needs to be there. Nontoxic alternatives are not used because they do not exist or cannot exist. Otherwise, certainly, they would be used, right?

Of course, this is wrong, and the myth is incorrect. The universe of organic molecules alone is vast; a subset of molecules containing C, N, O, and S atoms is estimated to contain $10^{63}$ theoretical chemicals with a molecular weight under 1000 amu.[8] There are approximately only $10^5$ chemicals that have been synthesized and a large number are still used in commerce today.[9] Recognizing that this is only a small sliver of the possible chemicals, and there are still almost infinite potential replacements for our current commercial chemicals certainly provides many opportunities for innovation (as well as job security for chemists). But first, there needs to be a recognition and conscious awareness by all stakeholders that this challenge is achievable.

It is only by knowing that benign molecular design is possible that consumers and customers can begin to **demand** it. When people are no longer satisfied that their packaging is unnecessarily persistent, that their agricultural chemicals unnecessarily contain endocrine disruptors, or that their building materials unnecessarily contribute a large percentage to the greenhouse gas burden—*then* safer chemicals will have value in the marketplace. Only then will market forces help drive the change to a more healthful chemical infrastructure.

Sometimes these changes in the status quo need an initiator to start them. Sometimes these transformations, from the old way of doing things to the new way, need a catalyst to speed them up. Engaging policymakers in developing regulations that require makers and users of toxic substances to justify the use and attest to the lack of an alternative that is less hazardous could be that catalyst. These kinds of policies have ways of making CEOs, COOs, and CFOs aware of their potential vulnerability and liability and transforming the use of hazardous substances into a material issue for the corporation.

Thoughtful CEOs are already taking the lead on this effort. This was demonstrated by the many outstanding examples that exist throughout every sector of industry with just a few being highlighted in the previous chapter. However, these examples remain the exception rather than the rule. The science that we have outlined in this book must lead, but the science alone cannot bring safer chemicals to scale such that it benefits everyone around the world.

For whatever reason, chemists have been given the gift of being able to intellectually understand how to manipulate matter such that they can introduce new substances into the world. With every gift comes responsibility—the responsibility to ensure that the new substances that we create are safe for people and the planet. Chemists are certainly up to the task. It will just be the next in a long list of scientific miracles produced by the field of chemistry and we will continue to discover, design, invent ... but then we have to remember, **First Do No Harm**.[10]

## References


1. (a) Baxter, P. J.; Waldron, H. A. Chapter 4: The introduction and monitoring of occupational diseases. In: Waldron, H. A. (Ed.), *Occupational Health Practice*, 3rd ed., Elsevier, 1989, 41–72; (b) NIOSH. Mercury. https://www.cdc.gov/niosh/topics/mercury/ (accessed 10/11/2021).
2. Ceballos, D. M.; Fellows, K. M.; Evans, A. E.; Janulewicz, P. A.; et al. Perchloroethylene and dry cleaning: it's time to move the industry to safer alternatives. *Front Public Health*, 2021, **9**, 638082.
3. (a) Meharg, A. Science in culture. *Nature*, 2003, **423**(6941), 688–688; (b) Ball, P. William Morris made poisonous wallpaper. *Nature*, 2003; (c) Batty, P. *The Anatomy of Color: The Story of Heritage Paints & Pigments*. Thames & Hudson: London, UK, 2017.
4. (a) Pira, E.; Donato, F.; Maida, L.; Discalzi, G. Exposure to asbestos: past, present and future. *J Thorac Dis*, 2018, **10**(Suppl 2), S237–S245; (b) OSHA. Asbestos. https://www.osha.gov/asbestos (accessed 10/11/2021); (c) American Cancer Society. Asbestos and cancer risk. https://www.cancer.org/cancer/cancer-causes/asbestos.html (accessed 10/11/2021).
5. (a) Blaufox, M. D. Radioactive artifacts: historical sources of modern radium contamination. *Semin Nucl Med*, 1988, **18**(1), 46–64; (b) Gillmore, G. K.; Crockett, R.; Denman, T.; Flowers, A.; et al. Radium dial watches, a potentially hazardous legacy? *Environ Int*, 2012, **45**, 91–98.
6. (a) Wani, A. L.; Ara, A.; Usmani, J. A. Lead toxicity: a review. *Interdiscip Toxicol*, 2015, **8**(2), 55–64; (b) Hauptman, M.; Bruccoleri, R.; Woolf, A. D. An update on childhood lead poisoning. *Clin Pediatr Emerg Med*, 2017, **18**(3), 181–192.

7. Coish, P.; Brooks, B. W.; Gallagher, E. P.; Kavanagh, T. J.; Voutchkova-Kostal, A.; Zimmerman, J. B.; Anastas, P. T. Current status and future challenges in molecular design for reduced hazard. *Acs Sustain Chem Eng*, 2016, **4**(11), 5900–5906. http://dx.doi.org/10.1021/acssuschemeng.6b02089

8. Bohacek, R. S.; McMartin, C.; Guida, W. C. The art and practice of structure-based drug design: a molecular modeling perspective. *Med Res Rev*, 1996, **16**(1), 3–50.

9. (a) Bond, G. G.; Garny, V. Inventory and evaluation of publicly available sources of information on hazards and risks of industrial chemicals. *Toxicol Ind Health*, 2019, **35**(11–12), 738–751; (b) Institute of Medicine. *Identifying and Reducing Environmental Health Risks of Chemicals in Our Society: Workshop Summary*. The National Academies Press: Washington, DC, 2014, p. 179.

10. (a) Anastas, N. D. A Framework for Hazard Reduction through Green Chemistry, Ph.D. dissertation, University of Massachusetts Boston, 2008. (b) Anastas, N. D. Connecting toxicology and chemistry to ensure safer chemical design, *Green Chem.*, 2016, 18, 4325–4331. https://doi.org/10.1039/C6GC00758A

# Index